An Introduction to Bayesian Inference, Methods and Computation

Nick Heard

An Introduction to Bayesian Inference, Methods and Computation

 Springer

Nick Heard
Imperial College London
London, UK

ISBN 978-3-030-82810-3   ISBN 978-3-030-82808-0   (eBook)
https://doi.org/10.1007/978-3-030-82808-0

This Springer imprint is published by the registered company Springer Nature Switzerland AG
The registered company address is: Gewerbestrasse 11, 6330 Cham, Switzerland

Preface

The aim of writing this text was to provide a fast, accessible introduction to Bayesian statistical inference. The content is directed at postgraduate students with a background in a numerate discipline, including some experience in basic probability theory and statistical estimation. The text accompanies a module of the same name, *Bayesian Methods and Computation*, which forms part of the online *Master of Machine Learning and Data Science* degree programme at Imperial College London.

Starting from an introduction to the fundamentals of subjective probability, the course quickly advances to modelling principles, computational approaches and then advanced modelling techniques. Whilst this rapid development necessitates a light treatment of some advanced theoretical concepts, the benefit is to fast track the reader to an exciting wealth of modelling possibilities whilst still providing a key grounding in the fundamental principles.

To make possible this rapid transition from basic principles to advanced modelling, the text makes extensive use of the probabilistic programming language *Stan*, which is the product of a worldwide initiative to make Bayesian inference on user-defined statistical models more accessible. Stan is written in C++, meaning it is computationally fast and can be run in parallel, but the interface is modular and simple. The future of applied Bayesian inference arguably relies on the broadening development of such software platforms.

Chapter 1 introduces the core ideas of Bayesian reasoning: Decision-making under uncertainty, specifying subjective probabilities and utility functions and identifying optimal decisions as those which maximise expected utility. Prediction and estimation, the two core tasks in statistical inference, are shown to be special cases of this broader decision-making framework. The application-driven reader may choose to skip this chapter, although philosophically it sets the foundation for everything that follows.

Chapter 2 presents representation theorems which justify the prior $\times$ likelihood formulation synonymous with Bayesian probability models. Simply believing that unknown variables are exchangeable, meaning probability beliefs are invariant to relabelling of the variables, is sufficient to guarantee that construction must hold. The prior distribution distinguishes Bayesian inference from frequentist statistical methods, and several approaches to specifying prior distributions are discussed. The prior and likelihood construction leads naturally to consideration of the posterior distribution, including useful results on asymptotic consistency and normality which suggest large sample robustness to the choice of prior distribution.

Chapter 3 shows how graphical models can be used to specify dependencies in probability distributions. Graphical representations are most useful when the dependency structure is a primary target of inferential interest. Different types of graphical model are introduced, including belief networks and Markov networks, highlighting that the same graph structure can have different interpretations for different models.

Chapter 4 discusses parametric statistical models. Attention is focused on conjugate models, which present the most mathematically convenient parametric approximations of true, but possibly hard to specify, underlying beliefs. Although these models might appear relatively simplistic, later chapters will show how these basic models can form the basis of very flexible modelling frameworks.

Chapter 5 introduces the computational techniques which revolutionised the application of Bayesian statistical modelling, enabling routine performance of inferential tasks which had previously appeared infeasible. Relatively simple Markov chain Monte Carlo methods were at the heart of this development, and these are explained in some detail. A higher level description of Hamiltonian Monte Carlo methods is also provided, since these methods are becoming increasingly popular for performing simulation-based computational inference more efficiently. For high-dimensional inference problems, some useful analytic approximations are presented which sacrifice the theoretical accuracy of Monte Carlo methods for computational speed.

Chapter 6 discusses probabilistic programming languages specifically designed for easing some of the complexities of implementing Bayesian inferential methods. Particular attention is given to Stan, which has experienced rapid growth in deployment. Stan automates parallel Hamiltonian Monte Carlo sampling for statistical inference on any suitably specified Bayesian model on a continuous parameter space. In the subsequent chapters which introduce more advanced statistical models, Stan is used for demonstration wherever possible.

Chapter 7 is concerned with model checking. There are no expectations for subjective probability models to be correct, but it can still be useful to consider how well observed data appear to fit with an assumed model before making any further predictions using the same model assumptions; it may make sense to reconsider alternatives. Posterior predictive checking provides one framework for model checking in the Bayesian framework, and its application is easily demonstrated in Stan. For comparing rival models, Bayes factors are shown to be a well-calibrated statistic for quantifying evidence in favour for one model or the other, providing a vital Bayesian analogue to Neyman-Pearson likelihood ratios.

Chapter 8 presents the Bayesian linear model as the cornerstone of regression modelling. Extensions from the standard linear model to other basis functions such as polynomial and spline regression highlight the flexibility of this fundamental model structure. Further extensions to generalised linear models, such as logistic and Poisson regression, are demonstrated through implementation in Stan.

Chapter 9 characterises nonparametric models as more flexible parametric models with a potentially infinite number of parameters, distributing probability mass across larger function spaces. Dirichlet process and Polya tree models are presented as respective nonparametric models for discrete and continuous random probability measures. Partition models such as Bayesian histograms are also included in this class of models.

Chapter 10 covers nonparametric regression. Particular attention is given to Gaussian processes, which can be regarded as generalisations of the Bayes linear model. Spline models and partition models are also re-examined in this context.

Chapter 11 combines clustering and latent factor models. Both classes of model assume a latent underlying structure, which is either discrete or continuous, respectively. Finite and infinite mixture models are considered for clustering data into homogeneous groupings. Topic modelling of text and other unstructured data is considered as both a finite and infinite mixture problem. Finally, continuous latent factor models are presented as an extension of linear regression modelling, through the inclusion of unobserved covariates. Again, example Stan code is used to illustrate this class of models.

Throughout the text, there are exercises which should form an important component of following this course. Exercises which require access to a computer are indicated with a □ symbol; these become increasingly prevalent as the chapters progress, reflecting the transition within the text from laying fundamental principles to applied practice.

London, UK   Nick Heard
June 2021

The original version of the book was revised. Copyright page text has been updated. The correction to the book is available at https://doi.org/10.1007/978-3-030-82808-0_12.

Contents

Chapter 1
Uncertainty and Decisions

1.1 Subjective Uncertainty and Possibilities

1.1.1 Subjectivism

In the seminal work of de Finetti (see the English translation of de Finetti 2017), the central idea for the Bayesian paradigm is to address decision-making in the face of uncertainty from a subjective viewpoint. Given the same set of *uncertain* circumstances, two decision-makers could differ in the following ways:

- How *desirable* different potential outcomes might seem to them.
- How *likely* they consider the various outcomes to be.
- How they feel their actions might *affect* the eventual outcome.

The Bayesian decision-making paradigm is most easily viewed through the lens of an individual making choices ("decisions") in the face of (personal) uncertainty. For this reason, certain illustrative elements of this section will be purposefully written in the first person.

This decision-theoretic view of the Bayesian paradigm represents a mathematical ideal of how a *coherent* non-self-contradictory individual should aspire to behave. This is a non-trivial requirement, made easier with various mathematical formalisms which will be introduced in the *modelling* sections of this text. Whilst these formalisms might not exactly match my beliefs for specific decision problems, the aim is to present sufficiently many classes of models that one of them might adequately reflect my opinions up to some acceptable level of approximation.

Coherence is also the most that will be expected from a decision-maker; there will be no requirement for me to choose in any sense the *right* decisions from any perspective other than my own at that time. Everything within the paradigm is subjective, even apparently absolute concepts such as *truth*. Statements of certainty such as "The true value of the parameter is x" should be considered shorthand for "It is my understanding that the true value of the parameter is x". This might seem pedantic,

© The Author(s), under exclusive license to Springer Nature Switzerland AG 2021
N. Heard, *An Introduction to Bayesian Inference, Methods and Computation*,
https://doi.org/10.1007/978-3-030-82808-0_1

but crucially allows contradictions between individuals, and between perspectives and *reality*: the decision-making machinery will still function.

1.1.2 Subjective Uncertainty

There are numerous sources of individual uncertainty which can complicate decision-making. These could include:

- Events which have not yet happened, but might happen some time in the future
- Events which have happened which I have not yet learnt about
- Facts which may yet be undiscovered, such as the truth of some mathematical conjecture
- Facts which may have been discovered elsewhere, but remain unknown to me
- Facts which I have partially or completely forgotten

In the Bayesian paradigm, these and other sources of uncertainty are treated equally. If there are matters on which I am unsure, then these uncertainties must be acknowledged and incorporated into a rational decision process. Whether or not I perhaps *should* know them is immaterial.

1.1.3 Possible Outcomes and Events

Suppose I, the decision-maker, am interested in a currently unknown outcome ω, and believe that it will eventually assume a single realised value from an exhaustive set of possibilities Ω. When considering uncertain outcomes, the assumed set of possibilities will also be chosen subjectively, as illustrated in the following example.

Example 1.1 If rolling a die, I might understandably assume that the outcome will be in $\Omega = \{\boxdot, \boxdot, \boxdot, \boxdot, \boxdot, \boxdot\}$. Alternatively, I could take a more conservative viewpoint and extend the space of outcomes to include some unintended or potentially unforeseen outcomes; for example, $\Omega = \{$Dice roll does not take place, No valid outcome, $\boxdot, \boxdot, \boxdot, \boxdot, \boxdot, \boxdot\}$.

Neither viewpoint in Example 1.1 could irrefutably be said to be *right* or *wrong*. But if I am making a decision which I consider to be affected by the future outcome of the intended dice roll, I would possibly adopt different positions according to which set of possible outcomes I chose to focus on. The only requirement for Ω is that it should contain every outcome I currently conceive to be possible and meaningful to the decision problem under consideration.

Definition 1.1 (*Event*) An *event* is a subset of the possible outcomes. An event $E \subseteq \Omega$ is said to occur if and only if the realised outcome $\omega \in E$.

1.2 Decisions: Actions, Outcomes, Consequences

1.2.1 Elements of a Decision Problem

Definition 1.2 (*Decision problem*) Following Bernardo and Smith (1994), a *decision problem* will be composed of three elements:

1. An **action** a, to be chosen from a set $\mathscr{A}$ of possible actions.
2. An uncertain **outcome** ω, thought to lie within a set Ω of envisaged possible outcomes.
3. An identifiable **consequence**, assumed to lie within a set $\mathscr{C}$ of possible consequences, resulting from the combination of both the action taken and the ensuing outcome which occurs.

Axioms 1 $\mathscr{C}$ will be totally ordered, meaning there exists an ordering relation $\leq_{\mathscr{C}}$ on $\mathscr{C}$ such that for any pair of consequences $c_1, c_2 \in \mathscr{C}$, necessarily $c_1 \leq_{\mathscr{C}} c_2$ or $c_2 \leq_{\mathscr{C}} c_1$.

If both $c_1 \leq_{\mathscr{C}} c_2$ and $c_2 \leq_{\mathscr{C}} c_1$, then we write $c_1 =_{\mathscr{C}} c_2$. This provides definitions of (subjective) preference and indifference between consequences.

Remark 1.1 Crucially, the ordering $\leq_{\mathscr{C}}$ is assumed to be subjective; my perceived ordering of the different consequences must be allowed to differ from that of other decision-makers.

Definition 1.3 (*Preferences on consequences*) Suppose $c_1, c_2 \in \mathscr{C}$. If $c_1 \leq_{\mathscr{C}} c_2$ and $c_1 \neq_{\mathscr{C}} c_2$, then c_2 is said to be a *preferable consequence* to c_1, written $c_1 <_{\mathscr{C}} c_2$. If $c_1 =_{\mathscr{C}} c_2$, then I am *indifferent* between the two consequences.

Definition 1.4 (*Action*) An *action* defines a function which maps outcomes to consequences. For simplicity of presentation, until Section 1.5.1 the actions in $\mathscr{A}$ will be assumed to be *discrete*, meaning that each can be represented by a generic form $a = \{(E_1, c_1), (E_2, c_2), \ldots\}$, where $c_1, c_2, \ldots \in \mathscr{C}$, and $E_1, E_2, \ldots$ are referred to as *fundamental events* which form a partition of Ω, meaning $\Omega = \cup_i E_i$, $E_i \cap E_j = \emptyset$ for $i \neq j$. Then, for example, if I take action a, then I anticipate that any outcome $\omega \in E_1$ would lead to consequence c_1, and so on.

Remark 1.2 When actions are identified, in this way, by the perceived consequences they will lead to under different outcomes, they are subjective.

1.2.2 Preferences on Actions

Rational decision-making requires well-founded preferences between possible actions. Let $a, a' \in \mathscr{A}$ be two possible actions, which for illustration could be written as

$$a = \{(E_1, c_1), (E_2, c_2), \ldots\},$$
$$a' = \{(E'_1, c'_1), (E'_2, c'_2), \ldots\}.$$

The overall desirability of each action will depend entirely on the uncertainty surrounding the fundamental events $E_1, E_2, \ldots$ and $E'_1, E'_2, \ldots$ and the desirability of the corresponding consequences $c_1, c_2, \ldots$ and $c'_1, c'_2, \ldots$. This can be exploited in two ways, which will be developed in later sections:

1. If I innately prefer action a to a', then this preference can be used to quantify my beliefs about the uncertainty surrounding the fundamental events characterising each action. This will form the basis for eliciting *subjective probabilities* (see Sect. 1.3).
2. Reversing the same argument, once I have elicited my probabilities for certain events then these can be used to obtain preferences between corresponding actions through the principle of maximising *expected utility* (see Sect. 1.4.1).

Definition 1.5 (*Preferences on actions*) For actions $a, a' \in \mathscr{A}$, a subjective decision-maker regarding a <u>not</u> to be a *preferable action* to a' is written $a \le a'$. For actions $a, a' \in \mathscr{A}$, if both $a \le a'$ and $a' \le a$, then a and a' are said to be *equivalent actions*, written $a \sim a'$.

Axioms 2 Preferences on actions must be compatible with preferences on consequences. Let E, F be events such that $\emptyset \subseteq E \subseteq F \subseteq \Omega$, and let $c_1, c_2 \in \mathscr{C}$ such that $c_1 \le_\mathscr{C} c_2$. Then the following preference on actions must hold:

$$\{(F, c_1), (\overline{F}, c_2)\} \le \{(E, c_1), (\overline{E}, c_2)\}.$$

Remark 1.3 The two actions $\{(F, c_1), (\overline{F}, c_2)\}$ and $\{(E, c_1), (\overline{E}, c_2)\}$ only differ in the consequences anticipated from any $\omega \in \overline{E} \cap F$; that is, the event $\overline{E} \cap F$ would lead to a consequence of c_1 under the first action and c_2 under the second.

Remark 1.4 By Axiom 2, for $\emptyset \subseteq E \subseteq \Omega$ and $c_1, c_2 \in \mathscr{C}$, if $c_1 \le_\mathscr{C} c_2$ then

$$\{(\Omega, c_1)\} \le \{(E, c_1), (\overline{E}, c_2)\} \le \{(\Omega, c_2)\}.$$

That is, if consequence c_2 is preferable to consequence c_1, then I should prefer a strategy which guarantees a consequence c_2 against carrying any risk of exposure to consequence c_1 through the occurrence of event E. Similarly, rather than guaranteeing the lesser consequence c_1, I should prefer a strategy whereby the occurrence of event E will improve the consequence to c_2.

1.3 Subjective Probability

1.3.1 Standard Events

Central to the definition given by Bernardo and Smith (1994) for *subjective probability* is the abstract concept of a continuous-indexed family of *standard events*, denoted S_x for $x \in [0, 1]$. These standard events are constructed in relation to a hypothetical, abstract experiment, such that under the *classical* perspective of probability one would assign probability x to the standard event S_x occurring, for $0 \le x \le 1$.

As an illustrative example, consider the hypothetical spinning wheel depicted in Fig. 1.1. This wheel is assumed to have unit circumference and to be plain in colour apart from a shaded sector of arc length $x \in [0, 1]$, creating an angle of $2\pi x$ radians from a horizontal axis. A fixed needle is mounted above the wheel as shown. It could be imagined that the wheel is to be spun (perhaps vigorously) from some arbitrary starting orientation; when the wheel has come to rest, one observes whether the fixed needle is lying within the shaded area of arc length x.

For each $x \in [0, 1]$, define the corresponding standard event

$$S_x = \{\text{Needle lies in the shaded area of arc length} x\}.$$

Classical probability would assign probability

Fig. 1.1 A spinning wheel with unit circumference and a fixed needle to depict a class of standard events S_x indexed by the arc length parameter $x \in [0, 1]$

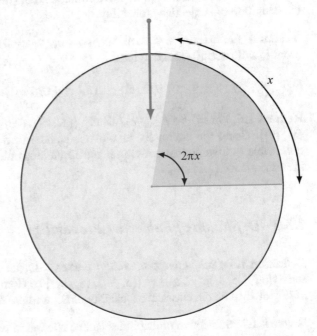

$$\frac{\text{arc length}}{\text{circumference}} = \frac{x}{1} = x$$

to the event S_x. Later, these standard events will be used to form the basis of a definition of subjective probability for quantifying individual uncertainty. Briefly, an individual will assign probability x to an event $E \subseteq \Omega$ if they would be indifferent between receiving a reward if E occurs or alternatively receiving the same reward if S_x occurs.

1.3.2   Equivalent Standard Events

Recall the standard events $\{S_x \mid 0 \le x \le 1\}$, introduced in Sect. 1.3.1.

Axioms 3 If $E \subseteq \Omega$, there exists a unique standard event S_x, $x \in [0, 1]$, such that for any $c_1, c_2 \in \mathscr{C}$ such that $c_1 <_\mathscr{C} c_2$,

$$\{(\overline{E}, c_1), (E, c_2)\} \sim \{(\overline{S_x}, c_1), (S_x, c_2)\}.$$

Remark 1.5 Axiom 3 uses the continuity in x of the collection of standard events $\{S_x : x \in [0, 1]\}$. It states that each event E can be mapped to a unique number $x \in [0, 1]$ through equivalence between E and the standard event S_x when imagined as alternative opportunities to improve consequences (from c_1 to c_2). This provides the definition of subjective probability.

Axioms 4 Let $c_1, c, c_2 \in \mathscr{C}$ such that $c_1 \le_\mathscr{C} c \le_\mathscr{C} c_2$. Then there exists a standard event S_x with $x \in [0, 1]$ satisfying

$$\{(\Omega, c)\} \sim \{(\overline{S_x}, c_1), (S_x, c_2)\}. \tag{1.1}$$

Remark 1.6 For $c_1 \le_\mathscr{C} c \le_\mathscr{C} c_2$, clearly $\{(\Omega, c_1), (\emptyset, c_2)\} \le \{(\Omega, c)\} \le \{(\emptyset, c_1), (\Omega, c_2)\}$. Using the continuity in x of the standard events $\{S_x : x \in [0, 1]\}$, it is reasonable to assume that between $\emptyset$ and Ω there should exist an event satisfying the equivalence (1.1).

1.3.3   Definition of Subjective Probability

Definition 1.6 (*Subjective probability*) For $E \subseteq \Omega$, let S_x be the standard event satisfying $\{(\overline{E}, c_1), (E, c_2)\} \sim \{(\overline{S_x}, c_1), (S_x, c_2)\}$ (Axiom 3). Then define the *probability* of E to be the classical probability of S_x, written $\mathbb{P}(E) = x$.

Remark 1.7 Subjective probabilities can be elicited in two stages: First, a continuous family of hypothetical standard events are constructed by the decision-maker, such that for each $x \in [0, 1]$ there is a corresponding standard event S_x with classical probability x. Second, a probability $\mathbb{P}(E) \in [0, 1]$ is assigned to an uncertain event $E \subseteq \Omega$ of interest by identifying equal preference between the two dichotomies $\{(\overline{\overline{E}}, c_1), (E, c_2)\}$ and $\{(\overline{S_{\mathbb{P}(E)}}, c_1), (S_{\mathbb{P}(E)}, c_2)\}$.

Remark 1.8 In some circumstances, the subjective assessment of the range of possible outcomes and the probabilities of events within that range may vary according to which action is being considered; for example, the decision problem may be choosing to role either one or two dice, with corresponding consequences resulting from the outcome. This presents no contradiction to the above definition, but all subjective probabilities should be regarded as conditional probabilities which implicitly condition on a particular action.

For further reading, see Sect. 5.3 of Gelman and Hennig (2017) for a discussion of subjective Bayesian reasoning within an interesting, wider discussion on objectivity and subjectivity in science.

1.3.4 Contrast with Frequentist Probability

It is worth noting the contrast of Definition 1.6 with *frequentist probability*. Under the frequentist interpretation, there exists a single probability of event E occurring, equal to the long run relative frequency at which E would occur in a potentially unlimited number of repetitions of the uncertain outcome.

Whilst these two interpretations of probability are fundamentally opposed, the two could easily coincide when subjective probabilities are determined by an individual using frequentist reasoning to arrive upon their own subjective beliefs.

1.3.5 Conditional Probability

Having started from an initial state of information, a decision-maker may need to update preferences and beliefs when additional information becomes available, encapsulated by the occurrence of some event $G \subset \Omega$. Such considerations require a notational extension for denoting consequently revised preferences on actions.

Definition 1.7 (*Conditional preferences on actions*) For actions $a, a' \in \mathscr{A}$, conditional preferences and equivalences assuming an event G has occurred will be denoted $a \leq_{|G} a'$ and $a \sim_{|G} a'$, respectively.

Using this notion of conditionally equivalent actions, Axiom 3 on equivalent standard events can be suitably extended.

Axioms 5 If $E, G \subseteq \Omega$, there exists a unique standard event S_x, $x \in [0, 1]$, such that for any $c_1, c_2 \in \mathscr{C}$ such that $c_1 < c_2$,

$$\{(\overline{E}, c_1), (E, c_2)\} \sim_{|G} \{(\overline{S_x}, c_1), (S_x, c_2)\}.$$

Remark 1.9 This axiom says that once we condition on an event G occurring, for any other event E we can still find an equivalent standard event.

Definition 1.8 (*Subjective conditional probability*) For $E, G \subseteq \Omega$, the *conditional probability* of E given G, written $\mathbb{P}_{|G}(E \mid G)$, is the index x of the standard event S_x satisfying $\{(\overline{E}, c_1), (E, c_2)\} \sim_{|G} \{(\overline{S_x}, c_1), (S_x, c_2)\}$.

Proposition 1.1 *For events $E, G \subseteq \Omega$ such that $\mathbb{P}(G) > 0$, the conditional probability of E given the assumed occurrence of G must necessarily be*

$$\mathbb{P}_{|G}(E \mid G) := \frac{\mathbb{P}(E \cap G)}{\mathbb{P}(G)}. \tag{1.2}$$

Proof See Sect. 1.4.5.

1.3.6 Updating Beliefs: Bayes Theorem

The updating Eq. (1.2) provides the unique recipe for how beliefs must be updated when additional information becomes available, and this can be further refined in the following theorem.

Theorem 1.1 (Bayes' theorem) *For events $E, G \subseteq \Omega$ such that $\mathbb{P}(G) > 0$,*

$$\mathbb{P}_{|G}(E \mid G) = \frac{\mathbb{P}_{|E}(G \mid E) \, \mathbb{P}(E)}{\mathbb{P}(G)}.$$

Proof From Proposition 1.1, $\mathbb{P}(E \cap G) = \mathbb{P}(G) \, \mathbb{P}_{|G}(E \mid G)$ and by symmetry, it must also hold that $\mathbb{P}(E \cap G) = \mathbb{P}(E) \, \mathbb{P}_{|E}(G \mid E)$. Hence $\mathbb{P}(G) \, \mathbb{P}_{|G}(E \mid G) = \mathbb{P}(E) \, \mathbb{P}_{|E}(G \mid E)$.

1.4 Utility

Definition 1.9 (*Utility function*) A *utility function* is a subjective, order-preserving mapping $u : \mathscr{C} \to \mathbb{R}$ such that $c_1 \leq_{\mathscr{C}} c_2 \iff u(c_1) \leq u(c_2)$.

Remark 1.10 A utility function assigns a subjective numerical value to each of the possible consequences.

Since each action-outcome pair (a, ω) in a decision problem leads to a consequence in $\mathscr{C}$, a utility function can equivalently be defined as a function $u : \mathscr{A} \times \Omega \to \mathbb{R}$, with

$$u(a, \omega) \equiv u(c)$$

for the corresponding consequence c for that action-outcome pair.

1.4.1 Principle of Maximising Expected Utility

In complex decision problems with uncertain outcomes, an additional principle on how to combine uncertainty with utilities is required to identify optimal decisions. This can be illustrated by a simple example.

Example 1.2 Consider two actions

$$a_1 = \{(\overline{E}, c_0), (E, c_1)\},$$
$$a_2 = \{(\overline{F}, c_0), (F, c_2)\},$$

for consequences $c_0 <_\mathscr{C} c_1 <_\mathscr{C} c_2$ and events $E, F \subset \Omega$ with $0 < \mathbb{P}(F) < \mathbb{P}(E) < 1$.

Without a method to trade-off between utility and uncertainty, there would be no basis on which to prefer either action. Action a_2 offers the opportunity of a superior consequence than a_1, but with lower enhancement probability.

This leads to the following axiom for preferences being determined by *expected utility*.

Definition 1.10 (*Expected utility of a deterministic action*) For a probability measure $\mathbb{P}$ and utility function u, the *expected utility* $\bar{u}(a)$ of an action $a = \{(E_1, c_1), (E_2, c_2), \ldots\} \in \mathscr{A}$ is defined to be

$$\bar{u}(a) := \sum_i \mathbb{P}(E_i)\, u(c_i).$$

Axioms 6 For two actions $a, a' \in \mathscr{A}$,

$$a \leq a' \iff \bar{u}(a) \leq \bar{u}(a'),$$

implying one action will be preferable to another if and only if it has higher expected utility.

Exercise 1.1 (*Linear transformations of utilities*) Show that decision problems are unaffected by positive-gradient linear transformations to the utility function.

Example 1.3 Continuing Example 1.2, but now assuming a utility function, the expected utilities of the two actions are

$$\bar{u}(a_1) = \{1 - \mathbb{P}(E)\}\, u(c_0) + \mathbb{P}(E)\, u(c_1),$$
$$\bar{u}(a_2) = \{1 - \mathbb{P}(F)\}\, u(c_0) + \mathbb{P}(F)\, u(c_2).$$

The action a_1 is preferable to a_2 if and only if

$$\frac{\mathbb{P}(E)}{\mathbb{P}(F)} > \frac{u(c_2) - u(c_0)}{u(c_1) - u(c_0)}.$$

The following two sections on bounded and unbounded decision problems together demonstrate that Axioms 4 and 6 ensure that the form of the utility function will be uniquely determined by the total ordering of $\mathscr{C}$, up to any positive-gradient linear rescaling (*cf.* Exercise 1.1).

1.4.2 Utilities for Bounded Decision Problems

Definition 1.11 (*Bounded decision problem*) A decision problem is said to be *bounded* if there exist worst and best consequences $c_*, c^* \in \mathscr{C}$ such that $\forall c \in \mathscr{C}$, $c_* \leq_\mathscr{C} c \leq_\mathscr{C} c^*$.

If the decision problem is bounded, then for simplicity and without loss of generality it can be assumed that $u(c_*) = 0, u(c^*) = 1$.

Then for any $c \in \mathscr{C}$, the order-preserving requirement of a utility function determines that $u(c) \in [0, 1]$ is the index of the standard event $S_{u(c)}$ such that $\{(\Omega, c)\} \sim \{(\overline{S_{u(c)}}, c_*), (S_{u(c)}, c^*)\}$ (Axiom 4).

Exercise 1.2 *Bounded utility.* Show that if $u(c_*) = 0$, $u(c^*) = 1$ and $c_* \leq_\mathscr{C} c \leq_\mathscr{C} c^*$, then necessarily by Axiom 6, $\{(\Omega, c)\} \sim \{(\overline{S_{u(c)}}, c_*), (S_{u(c)}, c^*)\}$.

1.4.3 Utilities for Unbounded Decision Problems

If the decision problem is not bounded, then for some $c_1 <_\mathscr{C} c_2$, (perhaps after linear rescaling) it could be assumed without loss of generality that $u(c_1) = 0, u(c_2) = 1$. Again, Axiom 4 and the order-preserving requirement then determine the rest of the utility function; specifically, for $c \in \mathscr{C}$:

1. If $c_1 \leq_\mathscr{C} c \leq_\mathscr{C} c_2$, $\{(\Omega, c)\} \sim \{(S_{u(c)}, c_2), (\overline{S_{u(c)}}, c_1)\}$.
2. If $c <_\mathscr{C} c_1$, then $u(c) < 0$ and if $\{(\Omega, c_1)\} \sim \{(\overline{S_x}, c), (S_x, c_2)\}$, then $u(c) = -x/(1-x)$.
3. If $c_2 <_\mathscr{C} c$, then $u(c) > 1$ and if $\{(\Omega, c_2)\} \sim \{(\overline{S_x}, c_1), (S_x, c)\}$, then $u(c) = 1/x$.

Exercise 1.3 *Unbounded utility.* Suppose $u(c_1) = 0, u(c_2) = 1$. Show that by Axiom 6, the following must hold.

(i) If $c <_{\mathscr{C}} c_1$ and $\{(\Omega, c_1)\} \sim \{(\overline{S_x}, c), (S_x, c_2)\}$, then $u(c) = -x/(1-x) < 0$.
(ii) If $c_2 <_{\mathscr{C}} c$ and $\{(\Omega, c_2)\} \sim \{(\overline{S_x}, c_1), (S_x, c)\}$, then $u(c) = 1/x > 1$.

Exercise 1.4 *Transitivity of preference.* Show that for $a, a', a'' \in \mathscr{A}$, if $a \leq a'$ and $a' \leq a''$, then $a \leq a''$.

Exercise 1.5 *Coherence with probabilities.* For events $E, F \subseteq \Omega$, show that if $\mathbb{P}(E) \leq \mathbb{P}(F)$ then $\{(F, c_1), (\overline{F}, c_2)\} \leq \{(E, c_1), (\overline{E}, c_2)\}$.

1.4.4 Randomised Strategies

Definition 1.12 *(Randomised action)* Let $G_1, G_2, \ldots$ be a partition of Ω. For each partition event G_i, suppose there is a corresponding action $a_{G_i} = \{(E_{i1}, c_{i1}), (E_{i2}, c_{i2}), \ldots\}$ which is determined to be taken if and only if G_i occurs. Denote this *randomised action* $a = \{(G_1, a_{G_1}), (G_2, a_{G_2}), \ldots\}$.

Remark 1.11 Randomised actions are a useful extension of the deterministic actions considered until now. Although sometimes counter-intuitive, in many circumstances they can sometimes be shown to correspond to optimal or near-optimal behaviours.

Definition 1.13 *(Expected utility of a randomised action)* The *expected utility of a randomised action* $a = \{(G_1, a_{G_1}), (G_2, a_{G_2}), \ldots\}$ is

$$\bar{u}(a) := \sum_i \mathbb{P}(G_i) \, \bar{u}_{|G_i}(a_{G_i} \mid G_i),$$

where $\bar{u}_{|G_i}(a_{G_i} \mid G_i) := \sum_j \mathbb{P}_{|G_i}(E_{ij} \mid G_i) \, u(c_{ij})$ is the conditional expected utility of action a_{G_i} given the occurrence of event G_i.

Remark 1.12 Definition 1.13 simply says that the expected utility of a randomised action is the expectation of the conditional expected utilities of the individual actions.

1.4.5 Conditional Probability as a Consequence of Coherence

By considering randomised actions, it can now be shown that the equation for conditional probability (1.2) is necessary when specifying subjective probabilities if those probabilities are to yield coherent expected utilities, and therefore coherent decisions.

Consider a randomised action $a = \{(G, a_G), (\overline{G}, a_{\overline{G}})\}$ such that $a_G = \{(E, c^*), (\overline{E}, c_*)\}$ and $a_{\overline{G}} = \{(\Omega, c_*)\}$, where $u(c_*) = 0$ and $u(c^*) = 1$. Then by Definition 1.13, a has expected utility

$$\bar{u}(a) = \mathbb{P}(G)\,\bar{u}_{|G}(a_G \mid G) + \mathbb{P}(\overline{G})\,\bar{u}_{|\overline{G}}(a_G \mid \overline{G})$$

$$= \mathbb{P}(G)[\mathbb{P}_{|G}(E \mid G)\,u(c^*) + \mathbb{P}_{|G}(\overline{E} \mid G)\,u(c_*)] + \mathbb{P}(\overline{G})\,\mathbb{P}_{|\overline{G}}(\Omega \mid \overline{G})\,u(c_*)$$

$$= \mathbb{P}(G)\,\mathbb{P}_{|G}(E \mid G).$$

But equivalently, a could be rewritten as a deterministic action, $a = \{(E \cap G, c^*), (\overline{E \cap G}, c_*)\}$. Then from Definition 1.10, it must also hold that

$$\bar{u}(a) = \mathbb{P}(E \cap G)\,u(c^*) + \mathbb{P}(\overline{E \cap G})\,u(c_*) = \mathbb{P}(E \cap G).$$

Hence for coherence in expected utility, $\mathbb{P}(E \cap G) = \mathbb{P}(G)\,\mathbb{P}_{|G}(E \mid G)$.

1.5 Estimation and Prediction

1.5.1 Continuous Random Variables and Decision Spaces

As noted in Definition 1.4, the initial notation used for actions has presumed discreteness, with a countable partition of Ω leading to countably many consequences and associated utilities. This section will consider cases where Ω and the space of actions might be uncountable.

Definition 1.14 (*Decision space*) A *decision space* (or continuous action space) is a set of mappings $\mathscr{D} = \{d : \Omega \to \mathscr{C}\}$ such that the consequence of taking a *decision* $d \in \mathscr{D}$ and observing outcome ω is $d(\omega) \in \mathscr{C}$.

Definition 1.15 (*Expected utility of a decision*) For a utility function $u : \mathscr{C} \to \mathbb{R}$, the *expected utility of a decision* $d : \Omega \to \mathscr{C}$ is the usual expectation

$$\bar{u}(d) = \int_{\Omega} u(d(\omega))\,\mathrm{d}\,\mathbb{P}(\omega).$$

Remark 1.13 If my probability distribution on Ω, $\mathbb{P}$, admits a density function representation f such that $\mathbb{P}(E) = \int_E f(\omega)\,\mathrm{d}\omega$ for all events $E \subseteq \Omega$, then

$$\bar{u}(d) = \int_{\Omega} u(d(\omega))\,f(\omega)\,\mathrm{d}\omega.$$

1.5.2 Estimation and Loss Functions

Consider the special case of the decision problem which is to estimate the future realised value of the unknown outcome $\omega \in \Omega$. In the typical notation of statistical estimation, a decision constitutes providing an estimated value $\hat{\omega}$. The eventual performance of that estimate is evaluated by a *loss function* $\ell(\cdot, \cdot)$, where $\ell(\hat{\omega}, \omega)$ quantifies an assumed penalty incurred by estimating the outcome with $\hat{\omega}$ when the true value transpires to be ω.

In this presentation of decision problems:

- The loss function value $\ell(\hat{\omega}, \omega)$ is the real-valued *consequence* of estimating ω with $\hat{\omega}$.
- The *utility* of the above consequence is simply the negative loss, $u(\ell(\hat{\omega}, \omega)) = -\ell(\hat{\omega}, \omega)$.

The decision of estimating ω by $\hat{\omega}$ could therefore be denoted

$$d_{\hat{\omega}} = \ell(\hat{\omega}, \cdot)$$

such that for $\omega \in \Omega$, $d_{\hat{\omega}}(\omega) = \ell(\hat{\omega}, \omega)$, and the expected utility is the negative expected loss,

$$\bar{u}(d_{\hat{\omega}}) = -\int_{\Omega} \ell(\hat{\omega}, \omega) \, f(\omega) \, d\omega.$$

Exercise 1.6 *Absolute loss (also known as L_1 loss).* If $\ell(\hat{\omega}, \omega) = |\hat{\omega} - \omega|$, show that the Bayes optimal decision is to estimate ω by the *median* of $\mathbb{P}$.

Exercise 1.7 *Squared loss (also known as L_2 loss).* If $\ell(\hat{\omega}, \omega) = (\hat{\omega} - \omega)^2$, show that the Bayes optimal decision is to estimate ω by the *mean* of $\mathbb{P}$.

Exercise 1.8 *Zero-one loss (also known as L_{∞} loss).* If $\ell(\hat{\omega}, \omega) = 1 - \mathbb{1}_{\{\hat{\omega}\}}(\omega)$, show that the Bayes optimal decision is to estimate ω by the *mode* of $\mathbb{P}$.

1.5.3 Prediction

In the preceding sections it would have been natural to envisage ω as a scalar number, such as the outcome from rolling a die. However, this need not be the case. *Bayesian prediction* is the task of estimating an entire probability distribution, rather than a scalar; correspondingly, in this case ω is an unknown probability distribution on a space X and Ω is a space of probability distributions on X which I believe contains ω.

As discussed throughout this chapter, in a Bayesian setting I will have my own beliefs about ω, characterised by my own subjective probability distribution $\mathbb{P}(E)$ for my probability that ω lies in a subset of probability distribution space $E \subseteq \Omega$.

To avoid self-contradiction, for *coherent* prediction it should be a requirement that my optimal decision when estimating ω should be to state my own beliefs. That is, the optimal decision $d_{\hat{\omega}}$ should satisfy, for events $F \subseteq X$,

$$\hat{\omega}(F) = \int_{\Omega} \omega(F) \, d\mathbb{P}(\omega), \tag{1.3}$$

where the right-hand side is my *marginal probability* for the event F, obtained as an expectation of the probability of F, $\omega(F)$, with respect to my uncertainty about ω encapsulated by $\mathbb{P}$.

Satisfying (1.3) clearly places constraints on what are allowable loss functions to lead to coherence. In fact, it can be shown (Bernardo and Smith 1994, Section 2.7) that the only *proper* loss functions for coherent prediction have a canonical form which is the well-known *Kullback-Leibler divergence* from information theory for measuring the difference of one probability distribution from another.

Definition 1.16 *Kullback-Leibler divergence* For two probability distributions P, Q where P is absolutely continuous with respect to Q, the *Kullback-Leibler divergence* (or simply, the *KL-divergence*) of Q from P is

$$\mathrm{KL}(P \parallel Q) := \int \log \frac{\mathrm{d}P}{\mathrm{d}Q} \, \mathrm{d}P = \mathbb{E}_P \, \log \frac{\mathrm{d}P}{\mathrm{d}Q}.$$

If p, q are corresponding density functions satisfying $p(x) > 0 \implies q(x) > 0$, then

$$\mathrm{KL}(p \parallel q) := \int p(x) \, \log \frac{p(x)}{q(x)} \, \mathrm{d}x = \mathbb{E}_p \, \log \frac{p(x)}{q(x)}.$$

Using this definition of KL-divergence, the necessary form for a proper loss function is

$$\ell(\hat{\omega}, \omega) = \mathrm{KL}(\omega \parallel \hat{\omega}) = \int_X \log \frac{\mathrm{d}\omega}{\mathrm{d}\hat{\omega}} \, \mathrm{d}\omega. \tag{1.4}$$

This justifies the use of KL-divergence for measuring discrepancy between two probability distributions from a Bayesian perspective.

Exercise 1.9 *KL-divergence non-negative.* For two probability density functions p, q, show that $\mathrm{KL}(p \parallel q) \geq 0$ with equality when $p = q$, and therefore KL-divergence is a proper loss function for prediction.

Chapter 2
Prior and Likelihood Representation

The first chapter introduced the philosophy of Bayesian statistics: when making individual decisions in the face of uncertainty, probability should be treated as a subjective measure of beliefs, where all quantities *unknown to the individual* should be treated as random quantities.

Eliciting individual probability assessments is a non-trivial endeavour. Even if I have a relatively well-formed opinion about some uncertain quantity, coherently assigning precise numerical values (probabilities) to all potential outcomes of interest for that quantity can be particularly challenging when there are infinitely many possible outcomes.

To counter these difficulties, it can be helpful to consider mathematical *models* to represent an individual's beliefs. There is no presumption that these models should be somehow *correct* in terms of representing true underlying dynamics; nonetheless, they can provide structure for representing beliefs coherently to a good enough degree of approximation to enable valid decision-making.

The main simplification which will be considered, *exchangeability*, occurs in contexts where a sequence of random quantities are to be observed and a joint probability distribution for the sequence is required. Symmetries in one's beliefs about sequences lead to familiar specifications of probability models which are often considered to be the hallmark of Bayesian thinking: a *likelihood* distribution combined with a *prior* distribution.

2.1 Exchangeability and Infinite Exchangeability

Let $X_1, X_2, \ldots$ be a sequence of real-valued random variables to be observed, which are mappings of an underlying unknown outcome $\omega \in \Omega$ with probability distribution $\mathbb{P}$.

The original version of this chapter has been revised due to typographic errors. The corrections to this chapter can be found at https://doi.org/10.1007/978-3-030-82808-0_12

© The Author(s), under exclusive license to Springer Nature Switzerland AG 2021, corrected publication 2022
N. Heard, *An Introduction to Bayesian Inference,*
Methods and Computation, https://doi.org/10.1007/978-3-030-82808-0_2

Definition 2.1 (*Exchangeability*) For $n \geq 1$, the finite sequence $X_1, \ldots, X_n$ is said to be *exchangeable* if, for any permutation σ on n symbols, their induced probability distribution satisfies

$$\mathbb{P}_{X_1, \ldots, X_n} = \mathbb{P}_{X_{\sigma(1)}, \ldots, X_{\sigma(n)}}$$

Definition 2.2 (*Infinite exchangeability*) An infinite sequence $X_1, X_2, \ldots$ is said to be *infinitely exchangeable* if, for all $n \geq 1$ and all choices of n indices $1 \leq i_1 < \ldots < i_n < \infty$, the subsequence $X_{i_1}, \ldots, X_{i_n}$ is exchangeable.

Remark 2.1 Exchangeability for a probability measure on n random variables simply implies that their probability distribution is invariant to the order in which they have been defined. Infinite exchangeability extends the concept to infinite sequences of random variables, requiring that any finite subsequence must be exchangeable.

Exchangeability can be a very natural (perhaps approximate) assumption in practical reasoning about uncertainty, such as assigning no importance to the order of observed outcomes from a (possibly unending) sequence of tosses of a coin.

2.2 De Finetti's Representation Theorem

On exchangeability, the Italian probability theorist Bruno de Finetti (1906–1985) is accredited with the following theorem which might be regarded as astonishing for its generality and impact.

Theorem 2.1 (De Finetti's representation theorem) *Let $X_1, X_2, \ldots$ be an infinitely exchangeable sequence of binary random variables, $X_i \in \{0, 1\}$. Then for any $n \geq 1$, there must exist a probability measure Q on $[0, 1]$ such that the corresponding mass function $\mathbb{p}_{X_1, \ldots, X_n}$ of the probability distribution $\mathbb{P}_{X_1, \ldots, X_n}$ satisfies*

$$\mathbb{p}_{X_1, \ldots, X_n}(x_1, \ldots, x_n) = \int_{\theta=0}^{1} \prod_{i=1}^{n} \theta^{x_i} (1 - \theta)^{1 - x_i} \, \mathrm{d}Q(\theta). \tag{2.1}$$

Proof See Bernardo and Smith (1994, p. 172).

Remark 2.2 Theorem 2.1 shows that any infinitely exchangeable sequence of binary random variables must arise as a sequence of independent and identically distributed Bernoulli(θ) random variables, with a single probability parameter θ drawn from some distribution Q.

The same property does not extend to finitely exchangeable sequences.

Exercise 2.1 (*Finitely exchangeable binary sequences*) Find an example of a finite sequence of binary random variables for which (2.1) does not hold.

Exercise 2.2 (*Predictive distribution for exchangeable binary sequences*) Suppose an infinitely exchangeable binary sequence $X_1, X_2, \ldots$. For $1 \leq m < n$, show that the conditional probability mass function for future elements $X_{m+1}, \ldots, X_n$ after observing $x_1, \ldots, x_m$ has the form

$$\mathbb{P}_{X_{m+1}, \ldots, X_n | x_1, \ldots, x_m}(x_{m+1}, \ldots, x_n) = \int_{\theta=0}^{1} \prod_{i=m+1}^{n} \theta^{x_i} (1-\theta)^{1-x_i} \, dQ(\theta \mid x_1, \ldots, x_m),$$

where

$$dQ(\theta \mid x_1, \ldots, x_m) = \frac{\prod_{i=1}^{m} \theta^{x_i} (1-\theta)^{1-x_i} \, dQ(\theta)}{\int_{\theta=0}^{1} \prod_{i=1}^{m} \theta^{x_i} (1-\theta)^{1-x_i} \, dQ(\theta)}.$$

Remark 2.3 Observing part of the sequence does not affect exchangeability, and therefore Theorem 2.1. The initial *prior* distribution $Q(\theta)$ is simply updated to the current *posterior* distribution $Q(\theta \mid x_1, \ldots, x_m)$.

Theorem 2.1 extends to non-binary, infinitely exchangeable sequences.

Theorem 2.2 *Let $X_1, X_2, \ldots$ be a sequence of real-valued random variables, $X_i \in \mathbb{R}$, which are believed to be infinitely exchangeable and let $\mathscr{R}$ be the space of all probability distributions on $\mathbb{R}$. Then for any $n \geq 1$, necessarily there exists a probability measure Q on $\mathscr{R}$ such that*

$$\mathbb{P}_{X_1, \ldots, X_n}(x_1, \ldots, x_n) = \int_{F \in \mathscr{R}} \prod_{i=1}^{n} F(x_i) \, dQ(F). \tag{2.2}$$

Proof See Bernardo and Smith (1994, p. 177).

Remark 2.4 From Bernardo and Smith (1994), the probability distribution Q has an operational interpretation, representing the uncertainty surrounding "what we believe the empirical distribution function would look like for a large sample".

Remark 2.5 In parametric statistical modelling, the probability function F in Theorem 2.2 is assumed to have a set parametric form $F(\cdot; \theta)$ for an unknown vector of parameters $\theta \in \mathbb{R}^k$. The representation then simplifies to

$$\mathbb{P}_{X_1, \ldots, X_n}(x_1, \ldots, x_n) = \int_{\theta \in \mathbb{R}^k} \prod_{i=1}^{n} F(x_i; \theta) \, dQ(\theta). \tag{2.3}$$

Similar to Exercise 2.2, the predictive distribution satisfies

$$\mathbb{P}_{X_{m+1}, \ldots, X_n | x_1, \ldots, x_m}(x_{m+1}, \ldots, x_n) = \int_{\theta \in \mathbb{R}^k} \prod_{i=m+1}^{n} F(x_i; \theta) \, dQ(\theta \mid x_1, \ldots, x_m),$$

where

$$dQ(\theta \mid x_1, \ldots, x_m) = \frac{\prod_{i=1}^{m} F(x_i; \theta) \, dQ(\theta)}{\int_{\theta=0}^{1} \prod_{i=1}^{m} F(x_i; \theta) \, dQ(\theta)}. \tag{2.4}$$

2.3   Prior, Likelihood and Posterior

Theorem 2.2 justifies the standard "prior $\times$ likelihood" approach commonly applied to Bayesian statistical modelling of real-valued data: assuming a sampling distribution comprising identically distributed observables which are conditionally independent given an unknown parameter, where the parameter is assumed to be an initial draw from some "prior" probability distribution.

The likelihood component in this mixture is common to both Bayesian and frequentist statistical approaches, and so more scepticism and attention is often directed towards how the prior component is specified in Bayesian methods. It is referred to as the "prior distribution" because it reflects one's beliefs about the generative mechanism, F, before observing any of the variables $X_1, X_2, \ldots$; in contrast the "posterior distribution" (2.4) reflects updated beliefs about F after observing those data.

2.3.1   Prior Elicitation

Eliciting the prior beliefs of an individual as a fully coherent probability distribution, obeying all the axioms of probability, presents a daunting challenge which has regularly been offered as a criticism of Bayesian reasoning. However, the difficulty in achieving this objective does not undermine its logical necessity. Rather than conceding defeat at the impossibility of exactly quantifying beliefs, various mathematical devices, such as exchangeability and different modelling ideas introduced in later chapters, are typically deployed to propose probability distributions which might hopefully reflect the degrees of belief of an individual to an acceptable degree of approximation.

2.3.2   Non-informative Priors

The difficulties of accurate prior elicitation for an individual, or perhaps a desire to identify decisions which might generalise to other individuals, often lead practitioners to try to propose *vague* or *non-informative* prior distributions, so that the observed data "may speak for themselves".

The word *vague* is often translated to mean *high variance*; assuming probability to be more widely spread around the mean will typically assign less mass to any one particular neighbourhood. However, without care this distinction may be artificial, as higher variance under one parameterisation may, under certain transformations, imply *lower* variance for a different parameterisation.

The word *non-informative* can assume a more specific interpretation: the prior distribution which would maximise the observed change between the prior and the corresponding posterior distribution, given observed data and a chosen distributional discrepancy measure. See Sect. 5.4 of Bernardo and Smith (1994) for discussion of *reference priors* and *reference decisions*; the latter identify the optimal decisions under a least informative prior—not for operational use for any individual decision-maker, but to serve as an illustrative benchmark for comparison.

Exercise 2.3 (*Variances under transformations*) Show that if $\theta \sim \text{Gamma}(a, b)$ (see Sect. A.3), then choices of (a, b) implying high variance for θ correspond to low variance for $1/\theta$.

2.3.3 Hyperpriors

For some applications, it can be convenient to specify a hierarchy of prior distributions. For example, it might seem easier for a practitioner to specify a prior distribution for a parameter θ conditional on the value of some other unknown parameter ϕ, $Q_{\theta|\phi}(\theta)$. A marginal prior for θ can then be recovered through specifying a prior measure for this *hyperparameter* (in frequentist statistics, *nuisance* parameter) ϕ, $Q_\phi(\phi)$, as then

$$Q_\theta(\theta) = \int_\phi Q_{\theta|\phi}(\theta) \; \mathrm{d}Q_\phi(\phi). \tag{2.5}$$

The additional level of prior modelling, $Q_\phi(\phi)$, is sometimes referred to as a *hyperprior*. By noting that a similar construction could be proposed for $Q_\phi(\phi)$, this hierarchical structure can be applied recursively to arbitrarily many nested levels.

2.3.4 Mixture Priors

A special case of (2.5) occurs when the hyperparameter ϕ is assumed to take one of only finitely many possible values; without loss of generality, suppose $\phi \in \{1, 2, \ldots, k\}$. Writing $w_i = \mathrm{d}Q_\phi(i)$, for $i = 1, \ldots, k$ with $\sum_{i=1}^k w_i = 1$, (2.5) simplifies to the *finite mixture prior*

$$Q_\theta(\theta) = \sum_{i=1}^k w_i \; Q_{\theta|\phi=i}(\theta)$$

on a finite collection of distributions $Q_{\theta|\phi=1}, \ldots, Q_{\theta|\phi=k}$.

Similarly, (2.5) is sometimes referred to as an *infinite mixture model*.

2.3.5 Bayesian Paradigm for Prior to Posterior Reporting

Given an initial probability distribution reflecting prior beliefs about F and then observing $X_1, \ldots, X_n$ as draws from F, Exercise 2.2 demonstrated the transition from prior distribution, through the likelihood function, to the posterior distribution (in this case for infinitely exchangeable random variables). This transformation was a simple application of Theorem 1.1, Bayes' theorem, and represents the only coherent mechanism for updating subjective probabilities.

In principle, the Bayesian paradigm for reporting scientific conclusions from a fixed collection of data suggests repeating this prior to posterior transformation for a range of different prior distributions, selected to cover a broad range of prior beliefs which may plausibly be held by the reader; for each prior distribution, the author would present the consequent posterior distribution and perhaps a corresponding optimal decision. However, in practice this procedure is often truncated, with authors preferring to show a single analysis under a non-informative prior (*cf.* Sect. 2.3.2), with the implication that inputting any more informative prior information would simply bias the conclusions in that direction, albeit by an unspecified amount.

2.3.6 Asymptotic Consistency

The sensitivity of posterior inferences to different prior distribution specifications (Sect. 2.3.5) is determined by the relative amount of information contained in the sample likelihood function. Suppose, as in (2.3), that $X_1, \ldots, X_n$ are assumed to be conditionally independent draws from the parametric distribution $F(\cdot; \theta)$ with a prior probability distribution $Q(\theta)$ for the unknown parameter θ. If the parametric form $F(\cdot; \theta)$ were *true*, and the true parameter which gave rise to these samples was θ^*, then the following proposition typifies several results which exist on posterior consistency.

Proposition 2.1 *Suppose $Q(\theta)$ is a discrete distribution with $\mathrm{d}Q(\theta^*) > 0$. If, for all other $\theta \neq \theta^*$ satisfying $\mathrm{d}Q(\theta) > 0$, $\mathrm{KL}(F(\cdot; \theta) \parallel F(\cdot; \theta^*)) > 0$, then*

$$\lim_{n \to \infty} \mathrm{d}Q(\theta \mid x_1, \ldots, x_n) = \mathbb{1}_{\{\theta^*\}}(\theta).$$

Proof See Bernardo and Smith (1994, p. 286).

Remark 2.6 The requirement $\mathrm{KL}(F(\cdot; \theta) \parallel F(\cdot; \theta^*)) > 0$ is sometimes referred to as *identifiability*. Under this condition, Proposition 2.1 states that as $n \to \infty$ the posterior distribution will converge to a single atom of mass located at the true value, provided that value had non-zero prior mass. In this sense, asymptotically the form of the prior $Q(\theta)$ does not matter beyond its *support*, $\{\theta : \mathrm{d}Q(\theta) > 0\}$.

Remark 2.7 It was remarked in Sect. 1.1.1 that the subjective Bayesian paradigm attaches no particular importance to absolute truths. From that perspective, Proposition 2.1 might appear to lack any operational significance; in subjective probability, there is no *true likelihood* and no *true parameter value*, nor will there be infinite random samples to observe.

However, there is a useful conclusion to draw: If you and I agree on exchangeability, the form of the sampling distribution $F(; \theta)$ and the range of values which θ reasonably might take, then even if we disagree on a form for the prior $Q(\theta)$, as we observe more data our posterior beliefs will uniformly converge. So for reporting scientific inference on "big data" applications, only the likelihood function really matters.

2.3.7 Asymptotic Normality

For continuous-valued parameters $\theta \in \mathbb{R}^k$, under some minor regularity conditions the posterior distribution is asymptotically normal, analogous to the result in classical statistics concerning the maximum likelihood estimator

$$\hat{\theta}_n = \arg\max_\theta \prod_{i=1}^n F(x_i; \theta). \tag{2.6}$$

For the maximum likelihood estimator, asymptotically $\hat{\theta}_n \sim \text{Normal}_k(\theta^*, I_n^{-1}(\theta^*))$, where $I_n(\theta)$ is the so-called *Fisher information matrix* of the likelihood function,

$$I_n(\theta) = -\frac{d^2}{d\theta^2} \sum_{i=1}^n \log F(x_i; \theta). \tag{2.7}$$

Proposition 2.2 *Let $m_0 = \arg\max_\theta dQ(\theta)$ be the mode of the prior distribution, and let $I_0(\theta) = -\frac{d^2}{d\theta^2} \log dQ(\theta)$. Then*

$$H_n = I_0(m_0) + I_n(\hat{\theta}_n) \tag{2.8}$$

is the posterior information matrix and

$$m_n = H_n^{-1}(I_0(m_0)m_0 + I_n(\hat{\theta}_n)\hat{\theta}_n) \tag{2.9}$$

the posterior mode, and asymptotically as $n \to \infty$,

$$Q(\theta \mid x_1, \ldots, x_n) \to \text{Normal}_k(\theta \mid m_n, H_n^{-1}) \to \text{Normal}_k(\theta \mid \theta^*, I_n^{-1}(\theta^*)).$$

Proof For a sketch proof involving a Taylor series expansion, see Bernardo and Smith (1994, p. 287).

Remark 2.8 Proposition 2.2 states that a large sample posterior distribution can be well approximated by a Gaussian; as $n \to \infty$ the mean of that Gaussian tends to the true value θ^* and the variance shrinks toward zero provided θ^* is identifiable, implying posterior consistency.

Exercise 2.4 *Asymptotic normality.* Let $x_1, \ldots, x_n$ be n observations from an infinitely exchangeable sequence of binary random variables as specified in Theorem 2.1. Suppose $Q(\theta) = \text{Beta}(\theta \mid a, b)$ (see Sect. A.2). Find the asymptotic normal distribution of θ as $n \to \infty$.

Chapter 3
Graphical Modelling and Hierarchical Models

In many contexts, straightforward exchangeability can be a useful simplifying assumption for specifying the joint probability distribution of random variables. But sometimes, an individual will require more complex structures of statistical dependence between random quantities to properly represent their beliefs. *Graphical models* provide a useful framework for characterising joint distributions for random variables, putting primary focus on characterising uncertainty in the dependency structure amongst the variables. Much of the material in this chapter is drawn from Barber (2012) and related resources.

Before introducing graphical models, some basic graph concepts and definitions are required to provide a language for relating probability distributions to graphs.

3.1 Graphs

3.1.1 Specifying a Graph

Definition 3.1 (*Graph*) A *graph* is a pair $\mathcal{G} = (V, E)$ where V is a non-empty set of entities, referred to as *nodes*, and $E \subset V \times V$ is a set of ordered pairs of nodes referred to as *edges*. The subset notation is strict, since for any $v \in V$ it is assumed here that $(v, v) \notin E$ (there are no self loops).

Definition 3.2 (*Directed and undirected graphs*) For a graph $\mathcal{G} = (V, E)$, if E is symmetric such that $(v, v') \in E \iff (v', v) \in E$, then the graph $\mathcal{G}$ is said to be *undirected*. Otherwise, the edges and graph are *directed*.

Fig. 3.1 An example graph with directed edges and the corresponding adjacency matrix

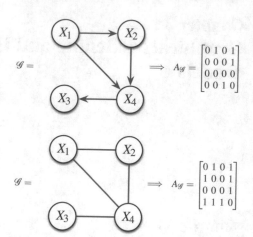

Fig. 3.2 An example graph with undirected edges and the corresponding adjacency matrix

Remark 3.1 Figures 3.1 and 3.2 provide diagrammatic examples of directed and undirected graphs. Each link drawn between nodes corresponds to an edge; in the directed graph, these links must have arrows to indicate their direction.

In the context of graphical modelling, the set of nodes in the graph will correspond to a finite set of random variables $V = \{X_1, \ldots, X_n\}$ for which a probability model must be constructed. The set of edges will correspond to proposed dependencies between variables, defined in different ways according to different modelling constructs.

Definition 3.3 (*Adjacency matrix*) For a finite graph $\mathscr{G} = (V, E)$, where $V = \{X_1, \ldots, X_n\}$, the *adjacency matrix* of the graph is a binary $n \times n$ matrix $A_{\mathscr{G}}$ with entries in $\{0, 1\}$, such that $(A_{\mathscr{G}})_{ij} = 1 \iff (X_i, X_j) \in E$.

Remark 3.2 An adjacency matrix provides an alternative characterisation of the edges in a graph. Figures 3.1 and 3.2 show the corresponding adjacency matrices implied by the example directed and undirected graphs. The diagonal elements will always be zero, and for an undirected graph the matrix is necessarily symmetric.

3.1.2 Neighbourhoods of Graph Nodes

Definition 3.4 (*Parents and children*) In a directed graph $\mathscr{G} = (V, E)$, the *parents* of node $X_i \in V$ is the set of nodes which connect to X_i through an edge in E, parents$(X_i) = \{X_j \in V : (X_j, X_i) \in E\}$. Similarly, the *children* of X_i is the subset of V connected to by X_i, children$(X_i) = \{X_j \in V : (X_i, X_j) \in E\}$.

Exercise 3.1 (*Identifying parents and children*) For the directed graph in Fig. 3.1, find the parents and children of each node in $V = \{X_1, X_2, X_3, X_4\}$.

Definition 3.5 (*Neighbours*) In an undirected graph $\mathcal{G} = (V, E)$, the *neighbours* of a node X_i, written neighbours(X_i), is simply the set of nodes in V connected to X_i by an edge in E, neighbours(X_i) = $\{X_j \in V : (X_i, X_j) \in E\}$.

Exercise 3.2 (*Identifying neighbours*) For the undirected graph in Fig. 3.2, find the neighbours of each node in $V = \{X_1, X_2, X_3, X_4\}$.

3.1.3 Paths, Cycles and Directed Acyclic Graphs

Definition 3.6 (*Path*) A sequence of distinct nodes $X_{i_1}, X_{i_2}, \ldots, X_{i_n}$ in V is a *directed path* in a graph $\mathcal{G} = (V, E)$ if, for each $1 \le j < n$, $(X_{i_j}, X_{i_{j+1}}) \in E$. The same sequence is an *undirected path* in the graph if, for each j, $(X_{i_j}, X_{i_{j+1}}) \in E$ or $(X_{i_{j+1}}, X_{i_j}) \in E$.

Definition 3.7 (*Cycle*) A *cycle* is a directed path $X_{i_1}, X_{i_2}, \ldots, X_{i_n}$ such that $X_{i_1} = X_{i_n}$.

Definition 3.8 (*Directed acyclic graph*) A *directed acyclic graph (DAG)* is a directed graph containing no cycles.

Remark 3.3 DAGs provide an important link between graph theory and probability modelling. In Sect. 3.2.1, they will be used to define a class of graphical models known as Bayesian belief networks. The direction of the links indicate an assumption of *causal* dependence. Figure 3.1 is an example of a DAG.

3.1.4 Cliques and Separation

Definition 3.9 (*Clique*) In an undirected graph $\mathcal{G} = (V, E)$, a *clique* is a fully connected subset of V. Furthermore, a clique is said to be *maximal* in the graph if there is no superset which is also a clique.

Exercise 3.3 (*Identifying cliques*) For the graph of Fig. 3.2, identify the maximal cliques.

Definition 3.10 (*Separation through a set*) For an undirected graph $\mathcal{G} = (V, E)$ and disjoint node subsets $\mathscr{A}, \mathscr{B}, \mathscr{C} \subset V = \{X_1, \ldots, X_n\}$, if every path from an element of $\mathscr{A}$ to an element of $\mathscr{B}$ contains an element of $\mathscr{C}$, then $\mathscr{C}$ is said to *separate* $\mathscr{A}$ from $\mathscr{B}$.

Exercise 3.4 (*Identifying separating sets*) For the graph in Fig. 3.2, find the separating sets.

Definition 3.11 (*Separation*) For $\mathscr{A}, \mathscr{B} \subset V$, $\mathscr{A}$ is *separated* from $\mathscr{B}$ in $\mathcal{G} = (V, E)$ if there is no path in $\mathcal{G}$ between an element of $\mathscr{A}$ and an element of $\mathscr{B}$.

3.2 Graphical Models

3.2.1 Belief Networks

Definition 3.12 (*Belief network*) Let $\mathscr{G}$ be a DAG on the node set of random variables $V = \{X_1, \ldots, X_n\}$. A *belief network* (also known as a *causal graph*) with graph $\mathscr{G}$ assumes the joint probability distribution factorises as

$$\mathbb{P}_{\mathscr{G}}(X_1, \ldots, X_n) = \prod_{i=1}^{n} \mathbb{P}(X_i \mid \text{parents}_{\mathscr{G}}(X_i)). \tag{3.1}$$

Remark 3.4 In a belief network, the set of DAG edges imply a collection of conditional independence statements, although not uniquely; one joint probability distribution can often be represented by multiple alternative DAGs.

Exercise 3.5 Interpreting the graph in Fig. 3.1 as a belief network, state the form of the implied joint probability distribution using the notation of (3.1).

3.2.1.1 Connectedness and Direct Separation

Definition 3.13 (*Connected graph*) A (directed or undirected) graph is said to be *connected* if there exists an undirected path between any two nodes in the graph.

Definition 3.14 (*Connected components*) If graph $\mathscr{G} = (V, E)$ is not connected, the nodes V can be uniquely partitioned into separated (see Definition 3.11) subsets $V_1, \ldots, V_k$, such that each subgraph $\mathscr{G}_i = (V_i, E \cap (V_i \times V_i))$ is connected and there are no connections in E between the subgraphs. The subgraphs $\mathscr{G}_1, \ldots, \mathscr{G}_k$ are said to be the *connected components* of $\mathscr{G}$.

Definition 3.15 (*Collider node*) In an undirected path within a directed graph, a node within the path is said to be a *collider* for that path if the edges on either side are both directed towards that node.

Exercise 3.6 (*Identifying colliders*) Figure 3.3 shows the three possible three-node paths that can exist within a directed graph. For each case, identify any colliders.

Definition 3.16 (*d-connected and d-separated*) Let $\mathscr{G} = (V, E)$ be a directed graph and $\mathscr{A}, \mathscr{B}, \mathscr{C} \subset V$ be disjoint node subsets.
    $\mathscr{A}$ is said to be *d-connected* to $\mathscr{B}$ by $\mathscr{C}$ if there exists an undirected path between an element of $\mathscr{A}$ and an element of $\mathscr{B}$ such that each element on the path is either

1. a non-collider which lies outside $\mathscr{C}$; or
2. a collider which either lies inside $\mathscr{C}$ or has a descendant in $\mathscr{C}$.

Otherwise, $\mathscr{C}$ is said to *d-separate* $\mathscr{A}$ from $\mathscr{B}$.

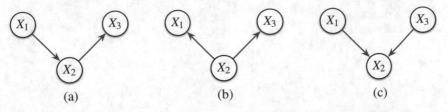

Fig. 3.3 The three possible directed graphs (up to label changes) with $|V| = 3$ and $|E| = 2$

Remark 3.5 The term d-separation is shorthand for "directional separation".

Remark 3.6 If $\emptyset$ d-separates $\mathscr{A}$ from $\mathscr{B}$, $\mathscr{A}$ and $\mathscr{B}$ are simply said to be d-*separated*.

Exercise 3.7 (*Identifying d-separated and d-connected nodes*) For each path in Fig. 3.3, identify any d-separated or d-connected nodes.

3.2.1.2 Independence and Conditional Independence

Proposition 3.1 *For a belief network on a directed graph $\mathscr{G} = (V, E)$ and disjoint node subsets $\mathscr{A}, \mathscr{B}, \mathscr{C} \subset V$, if $\mathscr{C}$ d-separates $\mathscr{A}$ from $\mathscr{B}$ then $\mathscr{A} \perp\!\!\!\perp \mathscr{B} \mid \mathscr{C}$ in the joint distribution $\mathbb{P}_{\mathscr{G}}$ of the belief network.*

Corollary 3.1 *Trivially from Proposition 3.1, the connected components of a graph in a belief network are independent.*

Exercise 3.8 (*Identifying conditional independencies in a belief network*) For each of the graphs in Fig. 3.3, state the dependence between X_1 and X_3 (i) marginally; (ii) conditionally given X_2.

3.2.2 Markov Networks

Definition 3.17 (*Markov network*) Let $\mathscr{G}$ be an undirected graph on the node set $\{X_1, \ldots, X_n\}$. A *Markov network* with graph $\mathscr{G}$ assumes the joint probability distribution factorises as

$$\mathbb{P}_{\mathscr{G}}(X_1, \ldots, X_n) \propto \prod_{i=1}^{C} \phi_i(\mathscr{X}_i), \tag{3.2}$$

where $\mathscr{X}_1, \ldots, \mathscr{X}_C$ are the maximal cliques of $\mathscr{G}$. The non-negative functions ϕ_i are sometimes referred to as *potentials*.

Exercise 3.9 (*Markov network distribution*) Interpreting the graph in Fig. 3.2 as a Markov network, state the form of the implied joint probability distribution using the notation of (3.2).

Fig. 3.4 A three-node graph
with two undirected edges

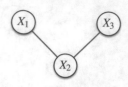

Definition 3.18 (*Pairwise Markov network*) Let $\mathscr{G} = (V, E)$ be an undirected graph on the node set $\{X_1, \ldots, X_n\}$. A *Markov network* with graph $\mathscr{G}$ assumes the joint probability distribution factorises as

$$\mathbb{P}_{\mathscr{G}}(X_1, \ldots, X_n) \propto \prod_{(X_i, X_j) \in E} \phi_{i,j}(X_i, X_j). \tag{3.3}$$

Exercise 3.10 (*Pairwise Markov network distribution*) Interpreting the graph in Fig. 3.2 as a pairwise Markov network, state the form of the implied joint probability distribution using the notation of (3.3).

Remark 3.7 The definitions of a Markov network and pairwise Markov network coincide if and only if the maximal cliques are all edges, meaning there are no triangles in the graph.

Remark 3.8 For the graph in Fig. 3.4, the definitions of Markov networks and pairwise Markov networks coincide, both implying $\mathbb{P}_{\mathscr{G}}(X_1, X_2, X_3) \propto \phi_1(X_1, X_2) \phi_2(X_2, X_3)$. In general, this simple graph would imply X_1 and X_3 are dependent, but conditionally independent given X_2.

3.2.2.1 Conditional Independence

Proposition 3.2 *Markov property. For disjoint $\mathscr{A}, \mathscr{B}, \mathscr{C} \subset V$, if $\mathscr{C}$ separates $\mathscr{A}$ from $\mathscr{B}$ in a graph $\mathscr{G} = (V, E)$ then $\mathscr{A} \perp\!\!\!\perp \mathscr{B} \mid \mathscr{C}$ in any Markov network on graph $\mathscr{G}$.*

Remark 3.9 In Fig. 3.4, $X_1 \perp\!\!\!\perp X_2 \mid X_3$. More generally, a node will be conditionally independent of any other nodes in the graph given its neighbours.

Definition 3.19 (*Markov random field*) Let $\mathscr{G}$ be an undirected graph on the node set $\{X_1, \ldots, X_n\}$. A *Markov random field* on $\mathscr{G}$ assumes the full conditional probability distributions satisfy

$$\mathbb{P}_{\mathscr{G}}(X_i \mid X_1, \ldots, X_{i-1}, X_{i+1}, \ldots, X_n) = \mathbb{P}_{\mathscr{G}}(X_i \mid \text{neighbours}_{\mathscr{G}}(X_i)).$$

Remark 3.10 The definition of a Markov random field is equivalent to the earlier definition of a Markov network, as characterised by (3.2).

Fig. 3.5 An example lattice graph

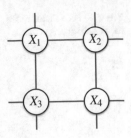

Exercise 3.11 Let $\mathcal{G} = (V, E)$ be a graph on $V = \{X_1, \ldots, X_n\}$. A multivariate normal distribution $N_n(\mu, \Sigma)$ is a *Gaussian Markov random field* (GMRF) with respect to $\mathcal{G}$ if the covariance matrix satisfies the condition

$$(\Sigma^{-1})_{ij} \neq 0 \iff (X_i, X_j) \in E.$$

Show that a GMRF satisfies Definition 3.19 for a Markov random field.

3.2.2.2 Lattice Models

Figure 3.5 shows an example of a *lattice* graph. Lattice graphs provide another case where the definitions of a Markov network/random field and a pairwise Markov network coincide. As Markov random fields, these structures are known as *lattice models*.

3.2.3 Factor Graphs

Factor graphs provide a further generalisation to (3.2), by allowing products of potentials (or *factors*) on arbitrary node subsets through the inclusion of additional (latent) *factor nodes* $\theta_1, \ldots, \theta_k$.

Definition 3.20 (*Factor graph*) Let $\mathcal{G}$ be an (undirected) graph on the extended node set $\{X_1, \ldots, X_n\} \cup \{\theta_1, \ldots, \theta_k\}$. A *factor graph* model assumes the joint probability distribution for $X_1, \ldots, X_n$ factorises as

$$\mathbb{P}_{\mathcal{G}}(X_1, \ldots, X_n) \propto \prod_{i=1}^{k} \phi_i(\text{neighbours}_{\mathcal{G}}(\theta_i) \cap \{X_1, \ldots, X_n\}). \qquad (3.4)$$

Remark 3.11 There should be no edges between factor nodes or variable nodes in a factor graph, since these would have no bearing on (3.4).

Fig. 3.6 As a factor model
for variables $X_1, \ldots, X_5$

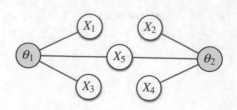

Remark 3.12 By introducing additional nodes, factor graphs can represent richer dependency structures than either belief networks or one Markov networks; one belief network or Markov network could correspond to multiple possible factor graphs.

Figure 3.6 shows an example factor graph, where the shaded nodes indicate latent factors.

The edge structure in Fig. 3.6 implies a factorisation of the joint distribution

$$\mathbb{P}_{\mathscr{G}}(X_1, X_2, X_3, X_4, X_5) \propto \phi_1(X_1, X_3, X_5)\phi_2(X_2, X_4, X_5).$$

3.2.3.1 Conditional Independence

Proposition 3.3 *For disjoint $\mathscr{A}, \mathscr{B}, \mathscr{C} \subset \{X_1, \ldots, X_n\}$, if $\mathscr{C}$ separates $\mathscr{A}$ from $\mathscr{B}$ in a factor graph $\mathscr{G} = (\{X_1, \ldots, X_n\} \cup \{\theta_1, \ldots, \theta_k\}, E)$ then $\mathscr{A} \perp\!\!\!\perp \mathscr{B} \mid \mathscr{C}$.*

Remark 3.13 In Fig. 3.6, $\{X_1, X_3\}$ and $\{X_2, X_4\}$ are conditionally independent given X_5.

3.3 Hierarchical Models

Section 2.3.3 introduced the idea of specifying probability distributions for unknowns through hierarchies. Such hierarchies can be interpreted as graphical models.

Definition 3.21 (*Hierarchical model*) A Bayesian *hierarchical model* for random variables $X_1, \ldots, X_n$ is a multiply-layered expression for the joint probability distribution with one or more *hyperparameters*.

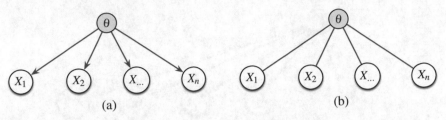

Fig. 3.7 Exchangeability for $X_1, \ldots, X_n$ represented as a **a** belief network or **b** factor graph

Remark 3.14 Hierarchical model formulations are equivalent to both belief networks (Sect. 3.2.1) and factor graphs (Sect. 3.2.3). They can be represented graphically in either way.

Example 3.1 De Finetti's representation Eqs. (2.2) and (2.3) for exchangeable variables $X_1, \ldots, X_n$ are simple hierarchical models. This representation is depicted graphically as both a belief network and a factor model in Fig. 3.7. The differences are the undirected edges and the explicit interpretation of θ as a latent parameter in the factor graph. The shaded nodes indicate latent random variables which will not be observed.

Example 3.2 Hierarchical models can be used to incorporate groups of dependencies into probability models. Again consider $X_1, \ldots, X_n$ to be exchangeable, but now suppose each X_i is a p-vector $X_i = (X_{i,1}, \ldots, X_{i,p})$ which can also be assumed to be exchangeable.

For example, on a degree course there could be n students who each sit p tests, such that $X_{ij} \in [0, 100]$ corresponds to the percentage score obtained by the ith student in the jth test. The implied $n \times p$ matrix (X_{ij}) could be regarded as a spreadsheet recording the student grades, where each row corresponds to a different student and each column to a different test.

Without further information about the students and the relative difficulty of the tests, a doubly-exchangeable assumption (for $X_1, \ldots, X_n$ and $X_{i,1}, \ldots, X_{i,p}$) could seem reasonable. (In contrast, assuming full exchangeability between all $n \times p$ test scores would be less comfortable, since each student might be expected to perform comparably in each of the different tests, according to their aptitude.)

Figure 3.8 shows the hierarchical model resulting from these two nested layers of exchangeability. The *root* node at the top of the hierarchy, F, is a probability distribution on the space of probability distributions. Each child node F_i is a draw from F corresponding to the grade probability distribution of student i. Finally, each individual test score X_{ij} is an independent random draw from the grade distribution F_i for that student.

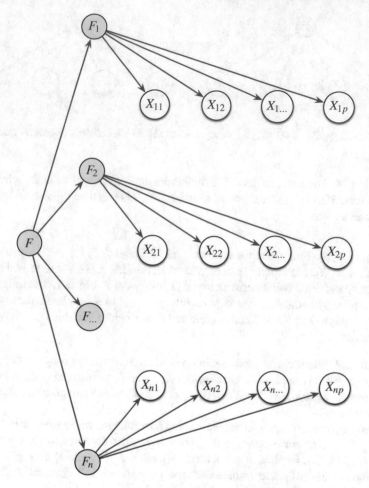

Fig. 3.8 A belief network representation of a hierarchical model for an $n \times p$ matrix of random variables (X_{ij}) with two layers of exchangeability: firstly in the rows, secondly in the row entries

Chapter 4
Parametric Models

This chapter introduces examples of parametric inferential models commonly used in the representation framework for exchangeable random variables from Chap. 2, and also the conditional distributions of more general dependency structures considered in Chap. 3.

4.1 Parametric Modelling

Suppose $\mathbf{x} = (x_1, \ldots, x_n)$ are the observed values of exchangeable random variables which are assumed to be conditionally independent given an unknown parameter $\theta \in \Theta$ (see Sect. 2.2). To simplify the notation of (2.3), the density of the joint distribution $\mathbb{P}_{X_1,\ldots,X_n}(x_1, \ldots, x_n)$ will now be written as $p(\mathbf{x})$; the prior density for θ, $\mathrm{d}Q(\theta)/\mathrm{d}\theta$ will be written simply as $p(\theta)$; $F(x; \theta)$ will be denoted $p(x \mid \theta)$; and the posterior density $\mathrm{d}Q(\theta \mid x_1, \ldots, x_n)/\mathrm{d}\theta$ will simply be written as $\pi(\theta)$. In this simplified notation, defining the joint likelihood

$$p(\mathbf{x} \mid \theta) := \prod_{i=1}^{n} p(x_i \mid \theta),$$

De Finetti's representation theorem becomes

$$p(\mathbf{x}) = \int_{\Theta} \prod_{i=1}^{n} p(x_i \mid \theta)\, p(\theta)\, \mathrm{d}\theta \tag{4.1}$$

© The Author(s), under exclusive license to Springer Nature Switzerland AG 2021
N. Heard, *An Introduction to Bayesian Inference, Methods and Computation*,
https://doi.org/10.1007/978-3-030-82808-0_4

and the posterior density for θ (2.4) can be expressed most simply as

$$\pi(\theta) \propto p(\mathbf{x} \mid \theta) \, p(\theta). \qquad (4.2)$$

Remark 4.1 In Bayesian inference it is common to see (posterior) probability densities being specified only up to a constant of proportionality. If $\pi(\theta) \propto g(\theta)$, then since all probability densities must integrate[1] to 1, it necessarily follows that $\pi(\theta) = g(\theta)/\{\int_\Theta g(\theta') \, d\theta'\}$. So (4.2) is simply a shortening of the full expression

$$\pi(\theta) = \frac{p(\mathbf{x} \mid \theta) \, p(\theta)}{\int_\Theta p(\mathbf{x} \mid \theta') \, p(\theta') \, d\theta'} = \frac{p(\mathbf{x} \mid \theta) \, p(\theta)}{p(\mathbf{x})}. \qquad (4.3)$$

However, a note of caution is required: Although the transition from an equation of proportionality (4.2) to equality (4.3) for the posterior density is automatic from a theoretical viewpoint, this normalisation requires evaluation of an integral in the denominator of (4.3) which may not always be analytically tractable.

4.2 Conjugate Models

Definition 4.1 (*Conjugacy*) A likelihood-prior representation (2.3) is said to be *conjugate* if the prior and posterior densities $p(\theta)$ and $\pi(\theta)$ from (4.2) are from the same parametric family.

Remark 4.2 For conjugacy to occur, the likelihood terms $p(x_i \mid \theta)$ must also resemble a density from the same parametric family as the prior $p(\theta)$, up to a constant of proportionality.

Tables 4.1 and 4.2 give examples of conjugate models for discrete and continuous random variables. In each row of either table, there is a likelihood model for which there exists a conjugate prior for one of the parameters, each time denoted θ. The right hand column shows the transformation from prior $p(\theta)$ to posterior $p(\theta \mid x)$ implied by a single observation x from the likelihood model $p(x \mid \theta)$.

Remark 4.3 In Table 4.1, the negative binomial distribution refers to the parameterisation

$$p(x \mid \theta) = \binom{r + x - 1}{r - 1} \theta^r (1 - \theta)^x,$$

corresponding to the number of "failures" observed before seeing r "successes" in a sequence of independent Bernoulli(θ) variables. Similarly the geometric distribution corresponds to Negative Binomial($1, \theta$), the distribution for the number of failures before the first success.

---

[1] Integration, here and elsewhere, refers to Lebesgue integration for densities of continuous random variables, and summation for densities (or mass functions) of discrete random variables.

Table 4.1 Conjugate parametric models for discrete random variables

Likelihood, $p(x \mid \theta)$	Conjugate prior, $p(\theta)$	Posterior, $p(\theta \mid x)$
Bernoulli(θ)	Beta(a, b)	Beta$(a + x, b + 1 - x)$
Geometric(θ)	Beta(a, b)	Beta$(a + 1, b + x)$
Binomial(m, θ)	Beta(a, b)	Beta$(a + x, b + m - x)$
Negative Binomial(r, θ)	Beta(a, b)	Beta$(a + r, b + x)$
Multinomial$_k(m, \theta)$	Dirichlet$_k(\alpha)$	Dirichlet$_k(\alpha + x)$
Poisson(θ)	Gamma(a, b)	Gamma$(a + x, b + 1)$

Table 4.2 Conjugate parametric models for continuous random variables

Likelihood, $p(x \mid \theta)$	Conjugate prior, $p(\theta)$	Posterior, $p(\theta \mid x)$
Uniform$(0, \theta)$	Pareto(a, b)	Pareto$(a + 1, \max\{b, x\})$
Exponential(θ)	Gamma(a, b)	Gamma$(a + 1, b + x)$
Gamma(ψ, θ)	Gamma(a, b)	Gamma$(a + \psi, b + x)$
Normal$_k(\theta, \Lambda^{-1})$	Normal$_k(\mu, P^{-1})$	Normal$_k((\Lambda + P)^{-1}(\Lambda x + P\mu), (\Lambda + P)^{-1})$
Normal$_k(\mu, \theta)$	Inverse Wishart$_k(a, b)$	Inverse Wishart$_k$ $(a + 1, b + (x - \mu) \cdot (x - \mu)^{\mathsf{T}})$

Generalisations of the results in Tables 4.1 and 4.2 to more than one observation from the likelihood model are straightforward; for a second observation, the posterior $p(\theta \mid x)$ from the right hand column adopts the role of the prior in the middle column, simply updating the parameter values within the same parametric family.

Exercise 4.1 Suppose $\mathbf{x} = (x_1, \ldots, x_n)$ are n independent Bernoulli(θ) random samples, and $\theta \sim$ Beta(a, b). Derive the posterior distribution for $\theta \mid \mathbf{x}$.

Exercise 4.2 Suppose $\mathbf{x} = (x_1, \ldots, x_n)$ are n independent Poisson(θ) random samples, and $\theta \sim$ Gamma(a, b). Derive the posterior distribution for $\theta \mid \mathbf{x}$.

Exercise 4.3 Suppose $\mathbf{x} = (x_1, \ldots, x_n)$ are n independent Uniform$(0, \theta)$ random samples, and $\theta \sim$ Pareto(a, b). Derive the posterior distribution for $\theta \mid \mathbf{x}$.

Exercise 4.4 Suppose $\mathbf{x} = (x_1, \ldots, x_n)$ are n independent Exponential(θ) random samples, and $\theta \sim$ Gamma(a, b). Derive the posterior distribution for $\theta \mid \mathbf{x}$.

Proposition 4.1 *For conjugate parametric models, the marginal likelihood $p(\mathbf{x})$ will have a closed-form equation.*

Proof For conjugate models the posterior density will have a closed analytic form; the marginal likelihood could therefore be obtained through rearranging (4.3),

$$p(\mathbf{x}) = \frac{p(\mathbf{x} \mid \theta) \, p(\theta)}{\pi(\theta)}. \tag{4.4}$$

Any terms involving θ in (4.4) will necessarily cancel, leaving a ratio of normalising constants from the likelihood and prior densities and the posterior density which do not depend on θ.

4.3   Exponential Families

Definition 4.2 (*Exponential family*) A density $p(x \mid \theta)$ belongs to an *exponential family* if there are functions g, h, η, T such that

$$p(x \mid \theta) = h(x)g(\theta) \exp\{\eta(\theta)^{\mathsf{T}} \cdot T(x)\}. \tag{4.5}$$

Proposition 4.2 *If $p(x \mid \theta)$ is an exponential family of the form* (4.5), *then any normalisable density satisfying*

$$p(\theta) \propto g(\theta)^r \exp(\eta(\theta)^{\mathsf{T}} \cdot s) \tag{4.6}$$

for $r > 0$ is a conjugate prior distribution for θ.

Proof From (4.2), the posterior density would be given up to proportionality by

$$p(\theta \mid x) \propto g(\theta)^{r+1} \exp\{\eta(\theta)^{\mathsf{T}} \cdot (s + T(x))\},$$

which is from the same parametric family as (4.6).

Remark 4.4 Proposition 4.2 provides another justification for the popularity of exponential family models in statistics; they all have conjugate Bayesian priors. All of the likelihood models in Tables 4.1 and 4.2 are exponential families.

4.4   Non-conjugate Models

For a given likelihood model in the De Finetti representation (4.1), adopting a conjugate prior distribution is certainly attractive for mathematical convenience. However, outside of simple exponential family examples, most likelihood models will not have a conjugate prior distribution.

Besides conjugate models, a tractable posterior distribution will always be theoretically available whenever the prior distribution is discrete and has finite support $\Theta = \{\theta : p(\theta) > 0\}$; in this case, the marginal likelihood $p(\mathbf{x})$ which serves as the normalising constant of (4.3) is simply the finite sum

$$p(\mathbf{x}) = \sum_{\theta \in \Theta} p(\mathbf{x} \mid \theta) \, p(\theta). \tag{4.7}$$

However, in practice, if the number of support points $|\Theta|$ is very large (for example, if θ is high-dimensional) then the summation (4.7) may still be too expensive to compute.

Moreover, a Bayesian model specification should aim to reflect subjective beliefs, and adopting a prior distribution simply for reasons of mathematical convenience is not consistent with this objective. Therefore, in many applications analysts will be faced with performing inference with non-conjugate statistical models with analytically intractable posterior distributions which can be identified only up to proportionality through (4.2).

4.5 Posterior Summaries for Parametric Models

Given a known posterior density $\pi(\theta)$ (4.2) obtained under an assumed parametric model, a decision-maker might be interested in visualising or quantifying some lower-dimensional summaries of this density; this can be particularly useful if the parameter θ is multi-dimensional, perhaps with high dimension, meaning the density $\pi(\theta)$ cannot simply be plotted.

4.5.1 Marginal Distributions

Suppose the parameter θ is a k-dimensional vector for $k > 1$, such that $\theta = (\theta_1, \ldots, \theta_k)$, and consider a decision problem to predict the value of a single component θ_j from that vector. In such circumstances the $(k-1)$-vector of remaining parameters, denoted

$$\theta_{-j} = (\theta_1, \ldots, \theta_{j-1}, \theta_{j+1}, \ldots, \theta_k), \tag{4.8}$$

are often referred to as *nuisance parameters*.

Recall from Sect. 1.5.3 that in the Bayesian paradigm, prediction corresponds to reporting one's subjective probability distribution for that parameter. Predicting the parameter component θ_j corresponds to reporting the *marginal posterior density* obtained from integrating out the nuisance parameters,

$$\pi(\theta_j) = \int_{\Theta_{-j}} \pi(\theta) \, d\theta_{-j}.$$

Exercise 4.5 Consider a bivariate target density function

$$\pi(\theta) = \frac{b^a \, \theta_1^a \, e^{-(b+\theta_2)\theta_1} \mathbb{1}_{[0,\infty)^2}(\theta)}{\Gamma(a)},$$

for $\theta = (\theta_1, \theta_2) \in [0, \infty)^2$ and constants $a, b > 0$. Calculate the marginal densities of θ_1 and θ_2. $[\Gamma(z) = \int_0^\infty x^{z-1} e^{-x} \, \mathrm{d}x.]$

4.5.2 Credible Regions

Alternatively, it may be of interest to identify a representative interval or region in which the parameter is believed to lie with some specified high probability, analogous to a *confidence interval* or region in frequentist statistics.

In subjective probability, the corresponding notion is referred to as a *credible region*.

Definition 4.3 (*Credible regions*) For $0 \le \alpha \le 1$, a $100\alpha\%$ *credible region* of a parameter θ with probability distribution $\mathbb{P}$ is a subset $R_\alpha \subseteq \Theta$ such that

$$\mathbb{P}(\theta \in R_\alpha) = \alpha.$$

Remark 4.5 For a given probability distribution and *coverage probability* $0 < \alpha < 1$, infinitely many valid credible intervals may exist.

For summarising a (marginal) posterior density $\pi(\theta)$ for a univariate continuous-valued parameter θ, a simple procedure for identifying a $100\alpha\%$ *credible interval* $[\theta_*, \theta^*]$ for θ is to identify interval boundaries such that

$$\int_{-\infty}^{\theta_*} \pi(\theta) \, \mathrm{d}\theta = \int_{\theta^*}^{\infty} \pi(\theta) \, \mathrm{d}\theta = \frac{1 - \alpha}{2}.$$

This particular choice of interval is sometimes referred to as an *equal-tailed credible interval*.

Exercise 4.6 Let $\pi(\theta) = \lambda e^{-\lambda\theta} \mathbb{1}_{[0,\infty)}(\theta)$ with $\lambda > 0$. For $0 \le \alpha \le 1$, calculate the equal-tailed $100\alpha\%$ credible interval for θ.

Chapter 5
Computational Inference

5.1 Intractable Integrals in Bayesian Inference

In Sect. 1.5, estimation and prediction were presented as Bayesian decision problems. Given a subjective probability distribution for an unknown quantity and a subjectively chosen utility or loss function, the Bayes estimate was shown to be the value which maximises expected utility or equivalently minimises expected loss. Obtaining this estimate apparently requires two stages of calculation: obtaining an analytic expression for the subjective probability distribution and then using this distribution to calculate expectations.

In the first stage, suppose I assume exchangeability for observable random variables $X_1, \ldots, X_n$ and a parametric representation (2.3) with an unknown parameter $\theta \in \Theta$. After observing values $x_1, \ldots, x_n$, my posterior distribution $\Pi(\theta)$ can theoretically be obtained through updating my prior beliefs via Bayes' theorem (4.3); however, the denominator of (4.3) is the result of a definite integral which was noted in Sect. 4.4 to be analytically intractable in many cases, leaving the posterior density only available up to an unknown constant of proportionality (4.2). Section 2.3.7 noted that almost any posterior distribution will asymptotically resemble a multivariate Gaussian, and in some large sample cases, this might provide an adequate approximation to the normalised posterior if the necessary maximum likelihood estimates can be calculated, but in general, these asymptotic arguments cannot be relied upon.

In the second stage, taking a *squared error* loss function as an example, it is understood from Sect. 1.5.2 that the Bayes estimate for θ under this loss function would be the *mean* value with respect to my (updated) subjective probability distribution, $\Pi(\theta)$, and the required calculation therefore requires a second integral

$$\mathbb{E}_\pi(\theta) := \int_\Theta \theta \, \pi(\theta) \, d\theta, \tag{5.1}$$

where $\pi(\theta) = d\Pi(\theta)/d\theta$ is the corresponding density function.

© The Author(s), under exclusive license to Springer Nature Switzerland AG 2021
N. Heard, *An Introduction to Bayesian Inference*,
Methods and Computation, https://doi.org/10.1007/978-3-030-82808-0_5

Slightly more generally, I might wish to estimate a transformation $g(\theta)$. Under squared error loss, the Bayes estimate would be the expectation of $g(\theta)$ with respect to my current beliefs about θ,

$$\mathbb{E}_\pi\{g(\theta)\} := \int_\Theta g(\theta)\,\pi(\theta)\,\mathrm{d}\theta. \tag{5.2}$$

Like the denominator of (4.3) when calculating a posterior distribution, in general, the integration required for calculating expectations (5.1) or (5.2) with respect to any *target density* $\pi(\theta)$ is also likely to be analytically intractable.

In summary, two sources of intractability have been identified:

1. Intractable posterior distribution calculation (4.3), where the normalising constant for $\pi(\theta)$ cannot be computed.
2. Intractable posterior expectation calculations (5.2), due to either the posterior $\pi(\theta)$ not being calculable (Item 1), or the integral of $g(\theta) \times \pi(\theta)$ not being tractable.

5.2 Monte Carlo Estimation

Monte Carlo methods are approximate, sampling-based approaches for evaluating expectations. They exploit the fact that if $M \geq 1$ random samples $\theta^{(1)}, \ldots, \theta^{(M)}$ can be obtained from the density π, then by linearity of expectation

$$\mathbb{E}_\pi\left\{\frac{1}{M}\sum_{i=1}^M g(\theta^{(i)})\right\} = \mathbb{E}_\pi\{g(\theta)\}.$$

Definition 5.1 (*Monte Carlo estimate of an expectation*) For samples $\theta^{(1)}, \ldots, \theta^{(M)}$ from π, the *Monte Carlo estimate (MC)* of $\mathbb{E}_\pi\{g(\theta)\}$ (5.2) is

$$\hat{\mathbb{E}}_\pi\{g(\theta)\} := \frac{1}{M}\sum_{i=1}^M g(\theta^{(i)}). \tag{5.3}$$

Remark 5.1 MC methods are well suited to addressing the case of Item 2 from Sect. 5.1, where a target density π might be fully known but the integrals required for calculating expectations with respect to π are not tractable.

Remark 5.2 As indicated above, by linearity of expectation (5.3) is an unbiased estimate of (5.2). By the strong law of large numbers, (5.3) converges to (5.2) almost surely.

Exercise 5.1 (*Monte Carlo probabilities*) Suppose $\theta^{(1)}, \ldots, \theta^{(M)}$ are random samples from a density $\pi(\theta)$ over Θ. State the Monte Carlo estimate of $\mathbb{P}_\pi(\theta \in A)$ for a region $A \subset \Theta$.

Exercise 5.2 (*Monte Carlo estimate of a conditional expectation*) Suppose $\theta^{(1)}, \ldots, \theta^{(M)}$ are random samples from a density $\pi(\theta)$ over Θ, $g(\theta)$ is a transformation of interest and $A \subseteq \Theta$. State a Monte Carlo estimate for the conditional expectation $\mathbb{E}_{\pi|A}(g(\theta) \mid \theta \in A)$.

Exercise 5.3 (*Monte Carlo credible interval*) For a univariate, real-valued parameter $\theta \in \mathbb{R}$, suppose $\theta^{(1)}, \ldots, \theta^{(M)}$ are random samples from a density $\pi(\theta)$ and $\theta_{(1)} \leq \ldots \leq \theta_{(M)}$ are the corresponding *order statistics*. For $0 \leq \alpha \leq 1$, use the order statistics to state a Monte Carlo approximated $100\alpha\%$ credible region for θ (*cf.* Sect. 4.5.2).

5.2.1 Standard Error

The standard error of an estimator is the standard deviation of the sampling distribution of the estimate, or more generally an estimate of that standard deviation.

Definition 5.2 (*Monte Carlo standard error*) For independent samples $\theta^{(1)}, \ldots,$ $\theta^{(M)} \sim \pi(\theta)$, the (estimated) standard error of the Monte Carlo estimate (5.3) is

$$\text{s.e.}\{\hat{\mathbb{E}}_\pi\{g(\theta)\}\} := \sqrt{\frac{1}{M(M-1)} \sum_{i=1}^{M} \left[g(\theta^{(i)}) - \hat{\mathbb{E}}_\pi\{g(\theta)\} \right]^2}. \tag{5.4}$$

Remark 5.3 (5.4) is useful for assessing convergence of the MC estimate (5.3) to (5.2). The standard error shrinks to zero at a rate proportional to $\sqrt{m}$.

5.2.2 Estimation Under a Loss Function

Suppose the quality of an estimate $\hat{\theta}$ of an unknown parameter θ is quantified by a loss function $\ell(\hat{\theta}, \theta)$. From Exercise 1.7, under a squared loss function $\ell(\hat{\theta}, \theta) = (\hat{\theta} - \theta)^2$ the optimal Bayesian estimate is known to correspond to the posterior mean for θ; in this case, the Monte Carlo estimate (5.3) would be directly applicable as the Bayes estimate for θ.

More generally, for an arbitrary loss function, the Bayes estimate may not take such a convenient form. However, by the same principle of minimising expected loss with respect to π, the Bayes estimate can still be identified in principle via Monte Carlo sampling by first using (5.3) to evaluate the expected loss for any proposed estimate $\hat{\theta}$,

$$\hat{\mathbb{E}}_\pi\{\ell(\hat{\theta}, \theta)\} = \frac{1}{M} \sum_{i=1}^{M} \ell(\hat{\theta}, \theta^{(i)}).$$

Second, the Bayes estimate is the value $\hat{\theta} \in \Theta$ which minimises the (estimated) expected loss,

$$\arg\min_{\hat{\theta} \in \Theta} \; \hat{\mathbb{E}}_\pi \{\ell(\hat{\theta}, \theta)\},$$

which will typically need to be identified through numerical optimisation.

□ **Exercise 5.4** (*Monte Carlo optimal decision estimation*) Suppose just three samples $\theta^{(1)} = 2$, $\theta^{(2)} = 5$, $\theta^{(3)} = 11$ are obtained from a target density $\pi(\theta)$ describing uncertainty about an unknown parameter θ. Assuming a Gaussian kernel loss function

$$\ell(\hat{\theta}, \theta) = -\exp\{-(\hat{\theta} - \theta)^2/10\},$$

plot the Monte Carlo expected loss function $\hat{\mathbb{E}}_\pi \{\ell(\hat{\theta}, \theta)\}$ for $\hat{\theta}$ over the interval $[0, 12]$ and numerically evaluate an approximate Bayes estimate of θ.

5.2.3 Importance Sampling

Sometimes, it may not be possible or convenient to draw random samples directly from $\pi(\theta)$ in order to calculate a Monte Carlo estimate, even when the density is fully known. *Importance sampling* generalises Monte Carlo estimation by supposing instead that samples $\theta^{(1)}, \ldots, \theta^{(M)}$ can be drawn from some other density $h(\theta)$; a weighted average of the corresponding values $g(\theta^{(1)}), \ldots, g(\theta^{(M)})$ is then taken to approximate (5.2), where the weights are chosen to precisely counterbalance the discrepancy between the sampling density h and the target density π.

For any density $h(\theta)$ which *dominates* $\pi(\theta)$, in the sense $\pi(\theta) > 0 \implies h(\theta) > 0$, the expectation (5.2) can be rewritten as

$$\mathbb{E}_\pi \{g(\theta)\} = \int_\Theta \left\{ \frac{g(\theta)\pi(\theta)}{h(\theta)} \right\} h(\theta) \, d\theta = \mathbb{E}_h \left\{ \frac{g(\theta)\pi(\theta)}{h(\theta)} \right\}. \tag{5.5}$$

Defining a so-called *importance* function as the ratio of the two densities, $w(\theta) = \pi(\theta)/h(\theta)$, the identity (5.5) implies

$$\mathbb{E}_\pi \{g(\theta)\} = \mathbb{E}_h \{w(\theta)g(\theta)\} \tag{5.6}$$

for any dominating density h, thereby expressing a general expectation with respect to π as a different expectation with respect to h. It immediately follows that a Monte Carlo approximation of (5.1) can be obtained using samples $\theta^{(1)}, \ldots, \theta^{(m)}$ drawn from h.

Definition 5.3 (*Importance sampling*) For samples $\theta^{(1)}, \ldots, \theta^{(M)}$ from h, the *importance sampling* Monte Carlo estimate of $\mathbb{E}_\pi\{g(\theta)\}$ (5.2), or equivalently (5.6), is

$$\hat{\mathbb{E}}_\pi^{\text{IS}}\{g(\theta)\} := \frac{1}{M} \sum_{i=1}^{M} w_i \, g(\theta^{(i)}), \tag{5.7}$$

where $w_i = w(\theta^{(i)}) = \pi(\theta^{(i)})/h(\theta^{(i)})$ are the *importance weights*.

Remark 5.4 Importance sampling Monte Carlo estimation with respect to π is equivalent to Monte Carlo estimation with respect to h,

$$\hat{\mathbb{E}}_\pi^{\text{IS}}\{g(\theta)\} = \hat{\mathbb{E}}_h\{w(\theta)g(\theta)\}. \tag{5.8}$$

Exercise 5.5 (*Importance sampling Monte Carlo standard error*) For independent samples $\theta^{(1)}, \ldots, \theta^{(M)} \sim h(\theta)$, state a formula for the standard error of the importance sampling Monte Carlo estimate $\hat{\mathbb{E}}_\pi^{\text{IS}}\{g(\theta)\}$ from (5.7).

Remark 5.5 The rate at which the importance sampling standard error from Exercise 5.5 shrinks to zero and the estimate (5.7) converges to the true value depends upon the functional ratio π/h. Good convergence can be obtained when h well approximates π and h possibly has heavier tails (Amaral Turkman et al. 2019).

5.2.4 Normalising Constant Estimation

Suppose $\pi(\theta)$ is known only up to proportionality by $\pi(\theta) \propto \gamma(\theta)$ for some known function γ, such that

$$\pi(\theta) = \gamma(\theta)/\gamma_*, \tag{5.9}$$

where $\gamma_* = \int \gamma(\theta) \, d\theta$.

Proposition 5.1 *Let $h(\theta)$ be a known density which dominates $\pi(\theta)$, such that h can be easily sampled from and let $\theta^{(1)}, \ldots, \theta^{(M)}$ be a random sample drawn from h. Then an importance sampling Monte Carlo estimate for the normalising constant γ_* is*

$$\hat{\gamma}_* = \frac{1}{M} \sum_{i=1}^{M} \frac{\gamma(\theta^{(i)})}{h(\theta^{(i)})}. \tag{5.10}$$

Proof This follows immediately from the identity (5.7).

5.2.4.1   Marginal Likelihood Estimation in Bayesian Inference

A simple application of estimating normalising constants occurs frequently within Bayesian inference, where the target density $\pi(\theta)$ is a posterior distribution known up to proportionality (4.2) by the product of two known functions, the likelihood and the prior,

$$\gamma(\theta) = p(\mathbf{x} \mid \theta) \, p(\theta).$$

The unknown normalising constant of (5.9) in this case is the *marginal likelihood*, $\gamma_* = p(\mathbf{x})$.

In the simplest implementation, the prior $p(\theta)$ could be used as the sampling density; given prior samples $\theta^{(1)}, \ldots, \theta^{(M)}$, the Monte Carlo estimate (5.10) of the normalising constant is

$$\hat{p}(\mathbf{x}) = \frac{1}{M} \sum_{i=1}^{M} p(\mathbf{x} \mid \theta^{(i)}). \tag{5.11}$$

Although sampling from the prior leads to a simplified equation for Monte Carlo estimation of the marginal likelihood, the standard error of (5.11) can be large if the likelihood is calculated on a large sample $\mathbf{x}$ which strongly outweighs the effects of the prior (*cf.* Sect. 2.3.6). As noted above, low variance estimates can be obtained when the sampling density closely resembles the target. Therefore, in large sample cases, a better importance sampling density could be the asymptotic normal distribution approximation of a posterior from Sect. 2.3.7.

5.3   Markov Chain Monte Carlo

If sampling directly from a particular target distribution $\Pi(\theta)$ (for the purpose of performing Monte Carlo integration) does not seem possible, and when it is not clear how to identify a suitable importance sampling density (*cf.* Sect. 5.2.3), Markov chain Monte Carlo (MCMC) methods provide a general solution for obtaining *approximate* samples from *any* target density. Conceptually, the idea is straightforward: A discrete-time homogeneous Markov chain of parameter values $\theta^{(1)}, \theta^{(2)}, \ldots$ is sampled according to a *transition probability density function* $p(\theta^{(i+1)} \mid \theta^{(i)})$, chosen such that the limiting (stationary) distribution of the parameter value sequence has density $\pi(\theta)$.

5.3.1   Technical Requirements of Markov Chains in MCMC

The following concepts of irreducibility, reversibility and stationarity are key to MCMC methods, described in more detail in Roberts and Rosenthal (2004). It should be supposed that an *initial value* $\theta^{(0)}$ is drawn from an initial probability distribution (possibly a point mass at some particular value) and then subsequent values $\theta^{(1)}, \theta^{(2)}, \ldots$ are drawn from the transition density $p(\theta^{(i+1)} \mid \theta^{(i)})$.

Definition 5.4 (*n-step transition probability distribution*) For $A \subseteq \Theta$ and $n \geq 1$, the n-step transition probability distribution, P^n, is the distribution of the state $\theta^{(n)}$ after n iterations of the Markov chain starting from $\theta^{(0)} \in \Theta$,

$$P^n(A \mid \theta^{(0)}) := \int_{\theta_n \in A} \int_{(\theta_1, \ldots, \theta_{n-1}) \in \Theta^{n-1}} p(\theta^{(1)} \mid \theta^{(0)}) \ldots p(\theta^{(n)} \mid \theta^{(n-1)}) \, d\theta_1 \ldots d\theta_n.$$

Definition 5.5 (*π-irreducible Markov chain*) A Markov chain with transition density $p(\theta^{(i+1)} \mid \theta^{(i)})$ is said to be π-*irreducible* if for each Π-measurable set $A \subset \Theta$ with $\Pi(A) > 0$ and for each $\theta \in \Theta$, there exists $n > 0$ such that $P^n(A \mid \theta) > 0$.

Remark 5.6 Informally, a π-irreducible Markov chain can eventually reach any neighbourhood of Θ where the target distribution has positive probability.

Definition 5.6 (*Aperiodic Markov chain*) A Markov chain with transition density $p(\theta^{(i+1)} \mid \theta^{(i)})$ is said to be *aperiodic* if, for each initial value $\theta^{(0)} \in \Theta$ and each Π-measurable set $A \subset \Theta$ with $\Pi(A) > 0$, $\{n \mid P^n(A \mid \theta^{(0)}) > 0\}$ has the greatest common divisor equal to 1.

Remark 5.7 Informally, an aperiodic Markov chain does not have a cyclic pattern to how it can arrive at different states.

Definition 5.7 (*π-reversible Markov chain*) A Markov chain transition density $p(\theta^{(i+1)} \mid \theta^{(i)})$ is said to be π-*reversible* if and only if

$$\pi(\theta) \, p(\theta' \mid \theta) = \pi(\theta') \, p(\theta \mid \theta'). \tag{5.12}$$

Remark 5.8 The condition (5.12) required for reversibility is sometimes referred to as *detailed balance*.

Definition 5.8 (*Stationary distribution*) The density $\pi(\theta)$ is said to be a *stationary distribution* for the transition density $p(\theta' \mid \theta)$ if and only if

$$\pi(\theta') = \int \pi(\theta) \, p(\theta' \mid \theta) \, d\theta.$$

Proposition 5.2 *If the transition density $p(\theta^{(i+1)} \mid \theta^{(i)})$ of a Markov chain satisfies detailed balance (is reversible) with respect to $\pi(\theta)$, then $\pi(\theta)$ is a stationary distribution.*

Proof

$$\int_\Theta \pi(\theta) \, p(\theta' \mid \theta) \, d\theta = \pi(\theta') \int_\Theta p(\theta \mid \theta') \, d\theta = \pi(\theta').$$

Remark 5.9 A consequence of Proposition 5.2 is that if an aperiodic Markov chain can be constructed which is irreducible and reversible with respect to a target density π, then samples from that Markov chain would eventually converge to be (dependent) samples from π.

In MCMC methods, a large number of samples are obtained from an aperiodic, π-irreducible, π-reversible Markov chain, perhaps discarding some initial *burn-in* samples before the chain is deemed to have sufficiently converged towards the target. The retained samples are treated as an approximate sample from π for the purposes of Monte Carlo estimation (Sect. 5.2). (The standard error formula 5.2 will not apply, even approximately, for MCMC samples since this was based on an assumption of independent samples.)

The next two sections introduce the mostly commonly used mechanisms for constructing a π-irreducible, π-reversible Markov chain required for MCMC: *Gibbs sampling* and the *Metropolis-Hastings algorithm*.

5.3.2 Gibbs Sampling

Suppose $\theta = (\theta_1, \ldots, \theta_k)$ is a k-vector of parameters with $k > 1$. Then, for $1 \leq j \leq k$, following (4.8) let θ_{-j} denote the $(k - 1)$-vector comprising the entries of θ with the jth component removed.

Gibbs sampling operates by selecting an index j (either randomly, or through a deterministic cycle) and sampling a new value for the component θ_j from the *full conditional* distribution

$$\pi(\theta_j \mid \theta_{-j}) := \frac{\pi(\theta)}{\pi(\theta_{-j})}, \tag{5.13}$$

where

$$\pi(\theta_{-j}) := \int_{\Theta_j} \pi(\theta) \, d\theta_j$$

is the marginal density for θ_{-j}.

Proposition 5.3 *A Markov chain with transition density*

$$p(\theta' \mid \theta) = \mathbb{1}_{\theta_{-j}}(\theta'_{-j})\pi(\theta'_j \mid \theta_{-j}) \tag{5.14}$$

is $\pi(\theta)$-reversible.

Proof Since $\theta'_{-j} = \theta_{-j}$ with probability 1 under (5.14), then for all such θ, θ',

$$\frac{\pi(\theta)}{\pi(\theta')} = \frac{\pi(\theta_j \mid \theta_{-j})}{\pi(\theta'_j \mid \theta'_{-j})} = \frac{p(\theta \mid \theta')}{p(\theta' \mid \theta)},$$

where the first equality derives from (5.13) and the second from (5.14).

Remark 5.10 Since the full conditional distributions are each π-reversible, a Markov chain which updates θ by successively sampling new component values from the full conditionals has stationary distribution π.

A cyclic implementation of the Gibbs sampling algorithm for obtaining approximate samples from π proceeds according to Algorithm 1.

Algorithm 1: Gibbs sampling

Result: M approximate samples from $\pi(\theta)$
1 Initialisation: Draw $\theta^{(0)} \in \Theta$ from an initial distribution;
2 **for** $i \leftarrow 1$ **to** M **do**
3 $\quad$ Set $\theta^{(i)} = \theta^{(i-1)}$ **for** $j \leftarrow 1$ **to** k **do**
4 $\quad\quad$ Draw $\theta_j^{(i)} \sim \pi(\theta_j^{(i)} \mid \theta_{-j}^{(i)})$ (see (5.13)) ;
5 $\quad$ **end**
6 **end**

Gibbs sampling can be particularly convenient within certainly classes of Bayesian hierarchical models (*cf.* Sect. 3.3); in such cases, the full conditional distributions can have tractable forms due to the hierarchical parameterisation. However, Gibbs sampling should be used with caution, particularly with high-dimensional (large k) models with strong dependencies between variables; in such cases the variances of the individual parameter full conditional distributions can become relatively small. Consequently, the sampler can fail to traverse multimodal target distributions, instead becoming stuck in *local modes*.

Exercise 5.6 (*Gibbs sampling*) Consider a mixture target distribution for $\theta = (\theta_1, \theta_2)^\mathsf{T}$ where θ_1, θ_2 are independent, identically normally distributed random variables with variance 1 and mean which is equal to μ with probability $\frac{1}{2}$ and equal to $-\mu$ otherwise, for some value $\mu > 0$.

The target density is depicted in Fig. 5.1 for two different values of the mean parameter, $\mu = 1$ and $\mu = 3$.

(i) State the target density $\pi(\theta_1, \theta_2)$ in terms of the standard normal density $\phi(z) = \frac{e^{-z^2/2}}{\sqrt{2\pi}}$.

(ii) Calculate the full conditional densities $\pi(\theta_1 \mid \theta_2)$ and $\pi(\theta_2 \mid \theta_1)$.

(iii) Show that Gibbs sampling will become less likely to move between two local modes as μ increases.

🖳 **Exercise 5.7** (*Gibbs sampling implementation*) Implement $M = 100$ iterations of Gibbs sampling (Algorithm 1) for the target distribution from Exercise 5.6, for the two cases (i) $\mu = 1$ and $\mu = 3$ depicted in Fig. 5.1. For each case, plot the trace of sampled values $\theta^{(1)}, \ldots, \theta^{(M)} \in \mathbb{R}^2$ to demonstrate the mixing of the Markov chain.

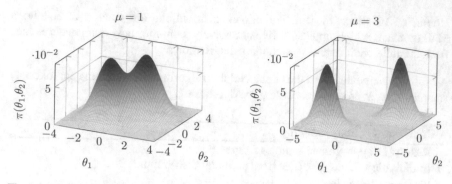

Fig. 5.1 Mixture density of two bivariate normal distributions with identity covariance matrix and means (μ, μ) and $(-\mu, -\mu)$

5.3.3 Metropolis-Hastings Algorithm

The Metropolis-Hastings algorithm provides a more general framework for constructing π-reversible Markov chains.

Let $q(\theta' \mid \theta)$ be the transition density of any irreducible Markov chain on Θ. Then the Metropolis-Hastings algorithm modifies the dynamics of this Markov chain by only *accepting* the moves proposed by q with probability

$$\alpha(\theta, \theta') = \min \left\{ 1, \frac{\pi(\theta')\, q(\theta \mid \theta')}{\pi(\theta)\, q(\theta' \mid \theta)} \right\} \tag{5.15}$$

and otherwise keeping the chain in its current state. The full algorithm is stated in Algorithm 2.

Algorithm 2: The Metropolis-Hastings algorithm

  Result: M approximate samples from $\pi(\theta)$
1 Initialisation: Draw $\theta^{(0)} \in \Theta$ from an initial distribution;
2 **for** $i \leftarrow 1$ **to** M **do**
3 $\quad$ Draw $\theta' \sim q(\theta' \mid \theta^{(i-1)})$;
4 $\quad$ Draw $u \sim \text{Uniform}(0, 1)$;
5 $\quad$ **if** $u < \alpha(\theta, \theta')$ *(see (5.15))* **then**
6 $\quad\quad \mid \ \theta^{(i)} = \theta'$;
7 $\quad$ **else**
8 $\quad\quad \mid \ \theta^{(i)} = \theta^{(i-1)}$;
9 $\quad$ **end**
10 **end**

Proposition 5.4 *The Markov chain transition density*

$$p(\theta' \mid \theta) = \alpha(\theta, \theta')q(\theta' \mid \theta) + \left(1 - \int_{\Theta} \alpha(\theta, \tilde{\theta})q(\tilde{\theta} \mid \theta) \, d\tilde{\theta}\right) \mathbb{1}_\theta(\theta') \qquad (5.16)$$

implied by the Metropolis-Hastings algorithm is π-reversible.

Exercise 5.8 (*Detailed balance of Metropolis-Hastings algorithm*) Prove Proposition 5.4 by checking the detailed balance equation (5.12) for the transition density (5.16), considering separately the two cases $\theta' = \theta$ and $\theta' \neq \theta$.

Remark 5.11 Since the transition density (5.16) is π-reversible, the Markov chain obtained from the Metropolis-Hastings algorithm has stationary distribution π.

Remark 5.12 The target density π only enters Algorithm 2 through the ratio $\pi(\theta)/\pi(\theta')$ in the acceptance probability (5.15); consequently, π (and also the proposal density q) only needs to be known up to proportionality to utilise the Metropolis-Hastings algorithm. This is a very useful property in Bayesian inference, where it has earlier been noted in Sect. 4.4 that a target posterior distribution can often only be identified up to an unknown normalising constant.

Remark 5.13 As with importance sampling (*cf.* Sect. 5.2.3), convergence of the Metropolis-Hastings algorithm depends upon the choice of the proposal density q, with good performance achieved when q closely resembles the target density π. The extreme case where $q(\theta' \mid \theta) = \pi(\theta')$ would lead to a sequence of independent samples drawn directly from π (all accepted with probability 1) and the algorithm reverts to straightforward Monte Carlo sampling (*cf.* Sect. 5.2).

Exercise 5.9 *Gibbs sampling as Metropolis-Hastings special case.* Show that Gibbs sampling (Sect. 5.3.2) is a special case of the Metropolis-Hastings algorithm with proposal density

$$q(\theta' \mid \theta) = \mathbb{1}_{\theta_{-j}}(\theta'_{-j})\pi(\theta'_j \mid \theta_{-j})$$

for updating the jth component of θ.

5.3.3.1 Random Walk

The most common implementations of the Metropolis-Hastings algorithm propose new values of θ using *local moves* generated by a simple random walk with a symmetric proposal density, such that $q(\theta' \mid \theta) = q(\theta \mid \theta')$. Under this symmetry, the Metropolis-Hastings acceptance probability (5.15) conveniently simplifies to the posterior ratio

$$\alpha(\theta, \theta') = \min\left\{1, \frac{\pi(\theta')}{\pi(\theta)}\right\}.$$

For example, in a univariate setting, commonly used symmetric proposals for local moves include

$$q(\theta' \mid \theta) \propto \exp(-(\theta' - \theta)^2/(2\varepsilon))$$

for a symmetric Gaussian proposal, or

$$q(\theta' \mid \theta) \propto \mathbb{1}_{(\theta-\varepsilon,\theta+\varepsilon)}(\theta')/(2\varepsilon)$$

for a symmetric uniform proposal. In either case, the parameter $\varepsilon > 0$ can be tuned to influence the *acceptance rate* of the proposed moves; as $\varepsilon \to 0$, the acceptance rate tends to 1, but at the expense of proper exploration of Θ. In practice, different values of ε can be explored to get a good trade-off between exploration and acceptance, with published research (Roberts et al. 1997) suggesting an acceptance ratio of 0.234 can optimise the efficiency of the algorithm under some quite general conditions. The consequent advice from the authors is to "tune the proposal variance so that the average acceptance rate is roughly 1/4".

Whilst a random walk Metropolis-Hastings algorithm avoids the difficulty of finding a proposal density that globally matches the target, these methods can sometimes perform poorly in practice by being slow to explore the parameter space, getting stuck in local modes of the target density. This phenomenon is sometimes described as *poor mixing*.

🖵 **Exercise 5.10** (*Metropolis-Hastings implementation*) Using a bivariate Gaussian proposal density

$$q(\theta' \mid \theta) \propto \exp\{-(\theta' - \theta)^\mathsf{T}(\theta' - \theta)/8\},$$

implement $M = 100$ iterations of the Metropolis-Hastings algorithm for the target distribution from Exercise 5.6. Address the two cases (i) $\mu = 1$ and $\mu = 3$ depicted in Fig. 5.1. For each case, plot the trace of sampled values $\theta^{(1)}, \ldots, \theta^{(M)} \in \mathbb{R}^2$ to demonstrate the mixing of the Markov chain.

5.4   Hamiltonian Markov Chain Monte Carlo

A more sophisticated implementation of the Metropolis-Hastings algorithm which can avoid the low acceptance rates or poor exploration of simplistic random walks is to generate proposals using dynamics inspired from Hamiltonian mechanics. The resulting algorithms are referred to as *Hamiltonian Monte Carlo* (HMC) methods (Neal 2011; Betancourt 2017).

To begin the mechanical analogy, the parameter θ of the target density $\pi(\theta)$ is first imagined to be the *location* of a body (typically a small ball) in a frictionless dynamical system. Second, the target density is augmented with a second parameter vector **p**, which acts as the *momentum* of the body. For elegance and simplicity, prior beliefs about the synthetic variable $\mathbf{p}$ are usually assumed to be described by a standard multivariate normal distribution that is statistically independent of θ; the joint density can then be written up to proportionality as

$$\tilde{\pi}(\theta, \mathbf{p}) \propto \pi(\theta) \, \exp(-\mathbf{p}^\mathsf{T}\mathbf{p}/2).$$

The negative logarithm of the augmented density (ignoring normalising constants) is assumed to correspond to total the energy of the body in this dynamical system, referred to as the *Hamiltonian*,

$$H(\theta, \mathbf{p}) := -\log \pi(\theta) + \frac{\mathbf{p}^\mathsf{T}\mathbf{p}}{2}. \qquad (5.17)$$

Continuing the mechanics analogy, the first term of (5.17) corresponds to the potential energy held by the body, proportional to the height of the body on a surface which has contours of $-\log \pi(\theta)$ at each location θ; and the second term corresponds to the kinetic energy of the body, proportional to the squared momentum, $\mathbf{p}^\mathsf{T}\mathbf{p}$. The lowest point on the surface, corresponding to minimal potential energy and therefore maximal kinetic energy, is the mode of the target density $\pi(\theta)$.

Returning to the Metropolis-Hastings algorithm, to propose new values in Θ from a current position, denoted here as $\theta(0)$, the idea is to consider a trajectory of the body through time after applying some momentum. Let $\theta(t)$ be the location of the body at time t, and $\mathbf{p}(t)$ the corresponding momentum. The principle of conservation of energy implies that when the extended target density is interpreted as the Hamiltonian of a closed dynamical system, the dynamics of that system should require $H(\theta, \mathbf{p})$ to be preserved. This leads to the *Hamiltonian equations* for the system:

$$\frac{d\theta(t)}{dt} = \frac{\partial H}{\partial \mathbf{p}}, \quad \frac{d\mathbf{p}(t)}{dt} = -\frac{\partial H}{\partial \theta}.$$

Evolving the extended parameters $\theta(t), \mathbf{p}(t)$ according to these equations would keep (5.17) constant, which corresponds to the body travelling along contours of the extended target density $\tilde{\pi}$. Therefore, proposing new θ values in approximate accordance with these dynamics can lead to proposals which are far away from the starting (previous) value but have similar target density $\pi(\theta)$, leading to good exploration and high acceptance rates.

In practice, the Hamiltonian dynamics are numerically approximated at interleaved time points using *leapfrog integration*; for an incremental time step $\varepsilon > 0$,

$$\mathbf{p}_j(t + \varepsilon/2) = \mathbf{p}_j(t) + (\varepsilon/2)\frac{\partial \log \pi(\theta(t))}{\partial \theta_j},$$

$$\theta_j(t + \varepsilon) = \theta_j(t) + \varepsilon \mathbf{p}_j(t + \varepsilon/2),$$

$$\mathbf{p}_j(t + \varepsilon) = \mathbf{p}_j(t + \varepsilon/2) + (\varepsilon/2)\frac{\partial \log \pi(\theta(t + \varepsilon))}{\partial \theta_j}. \qquad (5.18)$$

The partial derivatives of the target density are required in (5.18), implying the technique is only appropriate for continuous-valued parameters.

The Hamiltonian MCMC algorithm follows the Metropolis-Hastings algorithm (Algorithm 2), with proposal density $q(\theta' \mid \theta^{(i-1)})$ derived from a sampling procedure of first obtaining a new starting momentum $\mathbf{p}(0)$ from the standard multivariate normal, and second evolving the Hamiltonian dynamics implied by this initial momentum via the leapfrog algorithm for some number of time steps $L > 0$, beginning at $\theta(0) = \theta^{(i-1)}$. The algorithm for this proposal mechanism is given in Algorithm 3.

Algorithm 3: Hamiltonian Monte Carlo sampling

Result: A Metropolis-Hastings algorithm proposal $\theta' \mid \theta^{(i-1)}$

1 Initialisation: Set $\theta(0) = \theta^{(i-1)}$;
2 **for** $j \leftarrow 1$ **to** k **do**
3 $\quad$ Draw $\mathbf{p}_j(0) \sim \mathrm{N}(0, 1)$;
4 **end**
5 **for** $i \leftarrow 1$ **to** L **do**
6 $\quad$ Update $\mathbf{p}(t + (i - 1)\varepsilon), \theta(t + (i - 1)\varepsilon) \rightarrow \mathbf{p}(t + i\varepsilon), \theta(t + i\varepsilon)$ via (5.18);
7 **end**
8 Set proposal $\theta' = \theta(t + L\varepsilon)$.

5.5 Analytic Approximations

Markov chain Monte Carlo methods provide a general-purpose solution to computational Bayesian inference, providing estimates of arbitrarily high accuracy for any inference problem given sufficiently many iterations. However, in high-dimensional applications the time until reaching suitable convergence can be prohibitively long; in these circumstances, there is growing popularity for analytic approximate solutions. These approximations trade off the theoretic convergence guarantees of simulation-based inference methods for much faster inference procedures.

In this section, suppose the target density $\pi(\theta)$ corresponds to a posterior distribution density for a k-vector of parameters $\theta = (\theta_1, \ldots, \theta_k)$ after observing n likelihood samples $\mathbf{x} = (x_1, \ldots, x_n)$, such that π is known up to proportionality by (4.2).

5.5.1 Normal Approximation

Recall Proposition 2.2 from Sect. 2.3.7, which stated that for increasing sample sizes almost every target posterior distribution approaches an asymptotic normal distribution,

$$\pi(\theta) \to \text{Normal}_k(\theta \mid m_n, H_n^{-1}) \tag{5.19}$$

as $n \to \infty$, where m_n (2.9) and H_n (2.8) are respectively the posterior mode and information matrix. For approximate inference, this large sample property (5.19) can be exploited in several ways.

Most straightforwardly, the approximated normal distribution density could be directly substituted in place of the true target density $\pi(\theta)$, for example if this simplifies an expectation calculation (5.2). However, lower error approximations can be obtained using a so-called Laplace approximation.

5.5.2 Laplace Approximations

Combining (5.2) with the expression for the posterior distribution (4.3) obtained from Bayes' theorem, it follows that a posterior expectation for a function of interest $g(\theta)$ can be expressed as a ratio of two integrals,

$$\mathbb{E}\{g(\theta) \mid \mathbf{x}\} = \frac{\int_\Theta g(\theta)\, p(\mathbf{x} \mid \theta)\, p(\theta)\, d\theta}{\int_\Theta p(\mathbf{x} \mid \theta)\, p(\theta)\, d\theta}. \tag{5.20}$$

A *Laplace approximation* (Tierney and Kadane 1986) for (5.20) assumes a normal approximation to both the denominator and the numerator of this ratio. In general, the *Laplace method* of integration uses a second-order application of Taylor's theorem to approximate positive function integrands with normal distribution densities: let θ^* be the global maximum of a twice-differentiable function $h(\cdot)$, and $H(\cdot)$ the Hessian matrix of $h(\cdot)$, then

$$h(\theta) \approx h(\theta^*) + \frac{1}{2}(\theta - \theta^*)^\mathsf{T} H(\theta^*)^{-1}(\theta - \theta^*)$$

$$\implies \int e^{h(\theta)}\, d\theta \approx \frac{e^{h(\theta^*)}\,(2\pi)^{\frac{k}{2}}}{|-H(\theta^*)|^{\frac{1}{2}}}, \tag{5.21}$$

by comparison with the density of a normal distribution with mean vector θ^* and covariance matrix $-H(\theta^*)$. To apply Laplace's method to (5.20), it must be supposed that the function of interest $g(\theta)$ is positive almost everywhere. For the logarithm of the integrands in the denominator and numerator of (5.20), define

$$h(\theta) = \log p(\mathbf{x} \mid \theta) + \log p(\theta),$$

$$\tilde{h}(\theta) = \log g(\theta) + \log p(\mathbf{x} \mid \theta) + \log p(\theta).$$

The mode and Hessian of h are the posterior density mode and information matrix (m_n, H_n). Denoting the corresponding mode and Hessian of $\tilde{h}$ by $(\tilde{m}_n, \tilde{H}_n)$, the Laplace approximation of (5.20) by application of (5.21) is

$$\mathbb{E}\{g(\theta) \mid \mathbf{x}\} = \frac{\int_{\Theta} e^{\tilde{h}(\theta)} \, d\theta}{\int_{\Theta} e^{h(\theta)} \, d\theta} \approx \frac{|H_n|^{\frac{1}{2}} \, g(\tilde{m}_n) \, p_{\mathbf{x}|\theta}(\mathbf{x} \mid \tilde{m}_n) \, p_\theta(\tilde{m}_n)}{|\tilde{H}_n|^{\frac{1}{2}} \, p_{\mathbf{x}|\theta}(\mathbf{x} \mid m_n) \, p_\theta(m_n)}. \tag{5.22}$$

Remark 5.14 The Laplace approximation (5.22) is not invariant to transformations of the parameterisation θ. At least in principle, improved approximations can be achieved by using alternative parameterisations such that the resulting integrands in (5.20) more closely resemble normal distribution densities.

5.5.2.1 Approximating Marginal Distributions

Given a partition of the parameter k-vector $\theta = (\phi, \psi)$, such that ϕ is a k'-vector with $1 \leq k' < k$, a Laplace approximation can be used to approximate marginal distributions from the target density $\pi(\theta) \equiv \pi(\phi, \psi)$,

$$\pi(\phi) = \int_\Psi \pi(\phi, \psi) \, d\psi. \tag{5.23}$$

For a fixed value of ϕ, define

$$\tilde{h}_\phi(\psi) = \log p_{\mathbf{x}|\theta}(\mathbf{x} \mid \phi, \psi) + \log p_\theta(\phi, \psi).$$

Let $(\tilde{\psi}_{n,\phi}, \tilde{H}_{n,\phi})$ be the mode and Hessian of $\tilde{h}_\phi(\psi)$, again conditioning on the fixed value ϕ. Then using (5.21) in a similar way to deriving (5.22), a Laplace approximation for the marginal density (5.23) is

$$\pi(\phi) = \frac{\int_\Psi e^{\tilde{h}_\phi(\psi)} \, d\psi}{\int_\Theta e^{h(\theta)} \, d\theta} \approx \frac{|-H_n|^{\frac{1}{2}} \, p_{\mathbf{x}|\theta}(\mathbf{x} \mid \phi, \tilde{\psi}_{n,\phi}) \, p_\theta(\phi, \tilde{\psi}_{n,\phi})}{|-\tilde{H}_{n,\phi}|^{\frac{1}{2}} \, p_{\mathbf{x}|\theta}(\mathbf{x} \mid m_n) \, p_\theta(m_n) \, (2\pi)^{\frac{k-k'}{2}}}. \tag{5.24}$$

5.5.2.2 Integrated Nested Laplace Approximation

The integrated nested Laplace approximation (INLA), introduced by Rue et al. (2009), provides a particular useful implementation of Laplace approximations for an important model class known as *latent Gaussian models* (LGMs). An LGM has a likelihood function which assumes conditional independence given unobserved parameters θ and hyperparameters ϕ, such that the prior distribution for θ is a Gaussian Markov random field (GMRF, *cf.* Exercise 3.11). Accordingly,

$$x_i \sim p(x_i \mid \theta, \phi), \quad i = 1, \ldots, n,$$
$$\theta \mid \phi \sim \text{Normal}_k(0, \Sigma(\phi)),$$
$$\phi \sim p(\phi),$$

where $\Sigma(\phi)$ is a non-singular covariance matrix which can depend upon the hyper-parameter ϕ and whose inverse contains zeros according to the GMRF model. The normal distribution prior for θ makes models in this class well-suited to Laplace approximations. Noting the posterior density can be expressed as

$$\pi(\theta, \phi \mid \mathbf{x}) \propto p(\phi)|\Sigma(\phi)|^{\frac{1}{2}} \exp\left\{-\frac{1}{2}\theta^{\mathsf{T}} \Sigma^{-1}(\phi)\theta + \sum_{i=1}^{n} \log p(x_i \mid \theta, \phi)\right\},$$

the INLA approach combines multiple Laplace approximations for conditional distributions involving θ with numerical integration techniques for ϕ, and can therefore enable inference for problems with a very high dimensional θ parameter, provided ϕ has low dimension. In particular, using (5.24) the marginal posterior density for ϕ is approximated by

$$\hat{\pi}(\phi \mid \mathbf{x}) \propto \frac{p(\phi)|\Sigma(\phi)|^{\frac{1}{2}} \exp\left\{-\frac{1}{2}\tilde{\theta}_\phi^{\mathsf{T}} \Sigma^{-1}(\phi)\tilde{\theta}_\phi + \sum_{i=1}^{n} \log p(x_i \mid \tilde{\theta}_\phi, \phi)\right\}}{\hat{\pi}(\tilde{\theta}_\phi \mid \phi, \mathbf{x})},$$

where $\hat{\pi}(\theta \mid \phi, \mathbf{x})$ is the normal approximation to the corresponding full conditional distribution and $\tilde{\theta}_\phi$ is the constrained mode of that full conditional density for the fixed value ϕ.

Full details of the INLA method are beyond the scope of this text, but can be found in Rue et al. (2009). An open source implementation of the method is freely available, written in the statistical language R,[1] called R-INLA.[2]

5.5.3 *Variational Inference*

Not all posterior distribution densities can be well approximated with normal distributions, and so variational inference methods (Blei et al. 2017) explore alternative classes of approximating densities. Let $\mathscr{Q}$ be such a class of densities, referred to as the *variational family*. Then variational inference seeks to approximate the target density $\pi(\theta)$ with the closest member of the variational family, typically using Kullback-Leibler divergence (*cf.* Definition 1.16),

$$q^*(\theta) = \arg\min_{q \in \mathscr{Q}} \ \mathrm{KL}(q(\theta) \parallel \pi(\theta)). \tag{5.25}$$

The KL-divergence in (5.25) is taken in the reverse direction to the usual order, presented in (1.4), for comparing an estimated density with the truth. In (5.25), expectations are taken with respect to the estimating density q rather than the target

---

[1] https://www.r-project.org.

[2] https://www.r-inla.org.

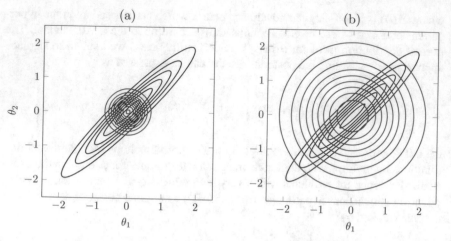

Fig. 5.2 Approximating a bivariate normal distribution with correlation .95, $\pi(\theta_1, \theta_2)$, with the closest independent bivariate normal distribution, $q(\theta_1, \theta_2)$, minimising **a** KL($q \parallel \pi$) or **b** KL($\pi \parallel q$)

π. This can lead to advantages in tractability, with freedom to chose a convenient form for the approximating density q.

An alternative algorithmic framework introduced by Minka (2001), known as *expectation propagation*, instead minimises the forward direction KL-divergence, KL($\pi(\theta) \parallel q(\theta)$). There are important differences between these two alternative formulations, which can be particularly important when approximating multimodal target distributions.

Specifically, KL($q(\theta) \parallel \pi(\theta)$) is more critical of discrepancies where $q(\theta)$ is large and $\pi(\theta)$ is small. Consequently, variational inference concentrates mass around a local mode of $\pi(\theta)$, and is said to be *zero-forcing* for q. In contrast, KL($\pi(\theta) \parallel q(\theta)$) is sensitive to $q(\theta)$ being small wherever $\pi(\theta)$ is large, and therefore must provide coverage to all modes of $\pi(\theta)$; it is therefore said to be *zero-avoiding* for q.

Following the example of Bishop (2006, p. 468), the two plots in Fig. 5.2 illustrate the contrasting approximations, obtained under the two alternative KL-divergence formulations, for a simple example where a bivariate normal distribution with correlation coefficient 0.95 is approximated by two independent univariate normal distributions. Both approximations correctly fit the mean of the target distribution, but the variational inference estimate in the left-hand plot focuses on the mode of the target distribution, where as the expectation-propagation approximation in the right-hand plot has higher variance, providing better coverage of the high target density region but also large areas of very low target density.

5.5.3.1 Evidence Lower Bound

With $\pi(\theta) \propto p(\mathbf{x}, \theta) = p(\mathbf{x} \mid \theta)p(\theta)$, minimising the KL-divergence (5.25) is equivalent to maximising the so-called *evidence lower bound*.

Definition 5.9 (*Evidence lower bound*) For fixed $\mathbf{x}$ and a probability density $q(\theta)$ satisfying $q(\theta) > 0 \implies p(\mathbf{x}, \theta) > 0$, the *evidence lower bound* (ELBO) is defined by

$$\text{ELBO}(q) := \mathbb{E}_q \log p(\mathbf{x}, \theta) - \mathbb{E}_q \log q(\theta). \tag{5.26}$$

Exercise 5.11 *ELBO equivalence.* Show that

$$\text{KL}(q(\theta) \parallel \pi(\theta)) = -\text{ELBO}(q) + \log p(\mathbf{x}), \tag{5.27}$$

and hence minimising $\text{KL}(q(\theta) \parallel \pi(\theta))$ is equivalent to maximising $\text{ELBO}(q)$.

Later in Sect. 7.1 which considers model uncertainty, the marginal likelihood $p(x)$ will be referred to as the *evidence* in favour of that particular probability model. This provides the reasoning behind the name of *evidence lower bound*: by (5.27),

$$\log p(\mathbf{x}) = \text{ELBO}(q) + \text{KL}(q(\theta) \parallel \pi(\theta))$$
$$\geq \text{ELBO}(q)$$

since KL-divergence is non-negative (*cf.* Exercise 1.9). Note the lower bound becomes an equality if $q = \pi$, corresponding to the approximation matching the target distribution.

Exercise 5.12 (*ELBO identity*) Show that $\text{ELBO}(q) = \mathbb{E}_q \log p(\mathbf{x} \mid \theta) - \text{KL}(q(\theta) \parallel p(\theta))$.

The specification of a variational inference method is completed by deciding upon the variational family $\mathcal{Q}$ over which (5.26) should be maximised. The approximating density (5.25) for π is then

$$q^*(\theta) = \arg\max_{q \in \mathcal{Q}} \ \text{ELBO}(q). \tag{5.28}$$

The computational difficulty of performing the optimisation (5.28) depends upon the complexity of the variational family. For tractable inference, the most common choice for $\mathcal{Q}$ is the so-called *mean-field variational family*.

Remark 5.15 To see the implicit trade-off implied by optimising the ELBO criterion, notice (5.26) is the sum of the expected log value, with respect to q, of the joint target density, $\pi(\mathbf{x}, \theta)$, plus a quantity referred to in information theory as the *entropy* of the approximating density, $-\mathbb{E}_q \log q(\theta)$. Without the entropy term, the maximisation would (in the limit) assign probability 1 to the posterior mode for θ;

however, that approximating density would have minimum entropy, and so instead the optimal q will distribute mass more widely around Θ, but still in areas where $\pi(\theta)$ is high.

5.5.3.2 Mean-Field Variational Inference

Definition 5.10 (*Mean-field variational family*) A *mean-field variational family* $\mathscr{Q}$ on Θ comprises probability density functions q with independent factors,

$$q(\theta) = \prod_{j=1}^{k} q_j(\theta_j). \tag{5.29}$$

Each factor q_j in (5.29) can assume a different parametric form, which might be necessary when some components of θ are unconstrained and continuous and others are possibly discrete. The assumption of independence implicit in (5.29) makes mean-field variational inference well-suited to optimisation using a technique called *coordinate ascent*. Once optimised, the mean-field variational estimate will take the same form,

$$q^*(\theta) = \prod_{j=1}^{k} q_j^*(\theta_j),$$

where $q_j^*(\theta_j)$ provides a *local variational approximation* of the marginal target density $\pi(\theta_j)$.

Remark 5.16 An unwanted consequence of the independence assumption of mean-field approximations is that component variances of $\pi(\theta)$ will typically be underestimated, as the main body of elongated, ellipsoidal covariance contours get approximated by smaller circles (*cf.* Fig. 5.2a).

5.5.3.3 Coordinate Ascent Variational Inference

Coordinate ascent algorithms discover local maxima of objective functions by sequentially optimising with respect to one parameter component whilst keeping the other components fixed. *Coordinate ascent variational inference* (CAVI) assumes a current mean-field approximation $q(\theta)$ (5.29), and then updates the jth component $q_j(\theta_j)$ to a locally optimal solution

$$q_j(\theta_j) \propto \exp\{\mathbb{E}_{q_{-j}} \log \pi(\theta_j \mid \theta_{-j})\} \propto \exp\{\mathbb{E}_{q_{-j}} \log p(\mathbf{x}, \theta)\},$$

where q_{-j} is the marginal density for θ_{-j} (4.8) of the current mean-field approximation,

$$q_{-j}(\theta_{-j}) = \prod_{\ell \neq j} q_\ell(\theta_\ell).$$

Exercise 5.13 (*CAVI derivation*) In coordinate ascent variational inference, show that

$$\arg\max_{q_j} \text{ELBO}(q) \propto \exp\{\mathbb{E}_{q_{-j}} \log \pi(\theta_j \mid \theta_{-j})\}.$$

The CAVI method is summarised in Algorithm 4. Each step of the algorithm maintains or increases the objective function $\text{ELBO}(q)$, and since $\text{ELBO}(q)$ is bounded above by $p(\mathbf{x})$, eventual convergence at a chosen tolerance threshold is guaranteed.

Algorithm 4: Coordinate ascent variational inference

Result: A mean-field variational estimate $q(\theta) = \prod_j q_j(\theta_j)$ for $\pi(\theta)$

1 Initialisation: Choose initial distributions $q_j(\theta_j)$, $j = 1, \ldots, k$, calculate $\text{ELBO}(\prod_j q_j)$ using (5.26);

2 **while** $\text{ELBO}(\prod_j q_j)$ *has not converged* **do**

3   **for** $j \leftarrow 1$ **to** k **do**

4   Set $q_j \propto \exp\{\mathbb{E}_{q_{\theta_{-j}}} \log p(\mathbf{x}, \theta)\}$;

5   **end**

6   Calculate $\text{ELBO}(\prod_j q_j)$ using (5.26)

7 **end**

☐ **Exercise 5.14** (*CAVI Gaussian approximation*) Suppose $\pi(\theta) = \text{Normal}_2(\theta \mid \mu, \Sigma)$ with $\mu \in \mathbb{R}^2$ and $\Sigma \in \mathbb{R}^{2 \times 2}$ positive-definite, and let $\mathcal{Q}_j = \{\text{Normal}(\theta_j \mid m, s^2) \mid m \in \mathbb{R}, s > 0\}$ be the variational family for component $j \in \{1, 2\}$ (*cf.* Fig. 5.2).

(i) Show that the CAVI algorithm local approximation for component j is $q_j(\theta_j) = \text{Normal}(\theta_j \mid m_j, s_j^2)$, where

$$m_j = \mu_j + \frac{\Sigma_{j\bar{j}}}{\Sigma_{\bar{j}\bar{j}}}(m_{\bar{j}} - \mu_{\bar{j}}), \quad s_j^2 = \Sigma_{jj} - \frac{\Sigma_{j\bar{j}}^2}{\Sigma_{\bar{j}\bar{j}}}$$

and $\bar{j} = 3 - j$ is the other component.

(ii) When will the algorithm converge?

(iii) Implement 200 iterations of the CAVI algorithm with target distribution mean $\mu = (0, 0)$, unit variances and correlation coefficient .95. Use starting values $m_1 = 2, m_2 = 3, s_1 = .1, s_2 = 9$. Make a contour plot of the target density and the mean-field variational approximation.

5.6 Further Topics

some of the fundamental computational methods for performing Bayesian inference. Much of the ongoing research activity in Bayesian statistics is focused in this area, with a broad range of sophisticated methods being developed. Further advanced topics include reversible jump Markov Monte Carlo, for transdimensional sampling from variable-dimension target distributions (Amaral Turkman et al. 2019, Chap. 7); sequential Monte Carlo sampling from a sequence of target distributions (Doucet et al. 2001); automatic differentiation for variational inference (Kucukelbir et al. 2017).

Such advanced techniques are well beyond the scope of this text, but the key principles introduced in this chapter provide the foundations for understanding the purpose behind these more advanced methods. The next chapter will illustrate computational packages which are openly available for users wishing to carry out different kinds of Bayesian analysis without addressing these research-level difficulties, where the complex sampling issues are kept "under the hood". Nonetheless, in diagnosing the performance of these (unavoidably imperfect) software tools it is important to possess this basic level of understanding of how their internal inferential processes operate.

Chapter 6
Bayesian Software Packages

The research-level complexity of performing Bayesian inference with the statistical models typically encountered in practical decision problems can provide a barrier to these methods being widely deployed. To alleviate this problem, a number of probabilistic programming languages have been developed specifically to automate Bayesian inference. This text will focus on the language *Stan*,[1] due to its widespread adoption and the depth of tutorial resources available. Brief details will also be given for two alternative libraries, *PyMC*[2] and *Edward*.[3] All three can be accessed through the general-purpose, interpreted programming language Python.[4]

To illustrate the use of computer software packages in performing Bayesian inference, the following hypothetical statistical model will be used to provide a working example.

6.1 Illustrative Statistical Model

Consider an example which further develops the graphical model structure presented in Example 3.2 and Fig. 3.8, which envisaged two layers of exchangeability for a hypothetical class of n students obtaining grades from p tests. Consider the following parametric model, consistent with Fig. 3.8, which assumes some exponential family distributions mentioned in Chap. 4:

---

[1] https://mc-stan.org
[2] https://docs.pymc.io
[3] http://edwardlib.org
[4] https://www.python.org

The original version of this chapter has been revised due to typographic errors. The corrections to this chapter can be found at https://doi.org/10.1007/978-3-030-82808-0_12

© The Author(s), under exclusive license to Springer Nature Switzerland AG 2021, corrected publication 2022
N. Heard, *An Introduction to Bayesian Inference*,
Methods and Computation, https://doi.org/10.1007/978-3-030-82808-0_2

$$\mu \sim \text{Normal}(0, \sigma^2/4)$$
$$\sigma^{-2} \sim \text{Gamma}(1, 1/2)$$
$$z_i \sim \text{Normal}(\mu, \sigma^2), \quad i = 1, \ldots, n$$
$$\theta_i = 1/\{1 + e^{-z_i}\}, \quad i = 1, \ldots, n$$
$$X_{ij} \sim \text{Binomial}(100, \theta_i), \quad i = 1, \ldots, n; \quad j = 1, \ldots, p. \tag{6.1}$$

Briefly, this model assumes a matrix, X, of student grades measured as integer percentage scores ranging between 0 and 100, such that the ith row corresponds to the ith student in a class. Each student grade X_{ij} is modelled by a binomial distribution with a student-specific probability parameter which is assumed to be the same for each test. This parameter is derived through a logistic transformation of an unobserved (*latent*), real-valued aptitude level z_i which is assumed to be normally distributed with unknown mean and variance which are assigned conjugate priors (*cf.* Sect. 4.2).

Below is some example Python code for simulating from this model, by default assuming thirty students sitting five tests.

```python
#! /usr/bin/env python
## student_grade_simulation.py

import numpy as np

def sample_student_grades(n=30,p=5,seed=0): #n students, p tests
    gen = np.random.default_rng(seed=seed)
    mu,sigma=sample_student_grade_parameters(gen)
    z=gen.normal(mu,sigma,size=n)
    X=[gen.binomial(100,1/(1.0+np.exp(-z_i)),size=p) for z_i in z]
    return(X,mu,sigma)

def sample_student_grade_parameters(gen,a=1,b=.5,m=0,tau=.5):
    sigma=1.0/np.sqrt(gen.gamma(a,1.0/b))
    mu=gen.normal(m,tau*sigma)
    return(mu,sigma)

#print(np.array(sample_student_grades()))
```

6.2 Stan

Stan, named after the Polish mathematician Stanislaw Ulam, is a probabilistic programming language written in C++ which uses sophisticated Monte Carlo and variational inference algorithms (see Chap. 5) for performing automated Bayesian inference. In particular, the default inferential method uses the *No-U-Turn Sampler* of Hoffman and Gelman (2014), which is an extension of Hamiltonian Monte Carlo (see Sect. 5.4). The derivatives required for performing HMC and other related inferential methods are calculated within Stan using automatic differentiation. The user simply has to declare the statistical model, import any data and then call a sampling routine.

Remark 6.1 Stan does not support sampling of discrete parameters, due to the reliance of the software on Hamiltonian Monte Carlo sampling methods. For problems involving discrete parameters, the Stan documentation recommends marginalising any discrete parameters where possible.

The following code (`student_grade_model.stan`) is written in the Stan language, declaring the statistical model (6.1) for student test grades from Sect. 6.1.

```
// student_grade_model.stan

data {
    int<lower=0> n; // number of students
    int<lower=0> p; // number of tests
    int<lower=0, upper=100> X[n, p]; // student test grades
    real<lower=0> tau;
    real<lower=0> a;
    real<lower=0> b;
}
parameters {
    real z[n];
    real mu;
    real<lower=0> sigma_sq;
}
transformed parameters {
    real<lower=0, upper=1> theta[n];
    real sigma;
    theta = inv_logit(z);
    sigma = sqrt(sigma_sq);
}
model {
    sigma_sq ~ inv_gamma(a,b);
    mu ~ normal(0, sigma * tau);
    z ~ normal(mu, sigma);
    for (i in 1:n)
        X[i] ~ binomial(100,theta[i]);
}
```

The `data{}` code block contains the quantities which are considered to be *known*. The quantities n (the number of students) and p (the number of tests) are declared as positive integers, and the student test scores (X_{ij}) are declared to be an $n \times p$ integer matrix taking values between 0 and 100. In the remainder of this block, the remaining required model hyperparameters and their constraints are listed.

The `parameters{}` code block declares the *unknown* quantities in (6.1): the n-vector of real number student aptitude values z_i, and the unknown mean and variance parameters (μ, σ^2) of the normal distribution for the z_i values.

The `transformed parameters{}` code block contains any parameter transformations which are helpful for stating the prior and likelihood models in the final `stan.model.sample{ }` block. In this case, the aptitude parameters z_i are converted to binomial parameters θ_i using the inverse logit function $\theta_i = 1/\{1 + e^{-z_i}\}$ as in (6.1), and also the aptitude standard deviation σ is obtained as the square root of the variance.

The `model{}` code block states the probability distributional assumptions from (6.1): An inverse-gamma distribution for σ^2; normal distributions for μ and the latent parameters z_i; and a binomial distribution for each individual percentage test score, using the student-specific transformed parameter θ_i.

6.2.1 PyStan

Stan can be accessed from a range of computing environments. In this text, it will be accessed using the Python interface, *PyStan*.[5] The following PyStan 3 code (`student_grade_inference_stan.py`) uses the Python simulation code and Stan model declaration code from above to simulate student test score data and then fit the underlying model to the data.

---

[5] https://pystan.readthedocs.io.

```python
#! /usr/bin/env python
## student_grade_inference_stan.py

import stan
import numpy as np
import matplotlib.pyplot as plt

# Simulate data
from student_grade_simulation import sample_student_grades
n, p = 30, 5
X, mu, sigma = sample_student_grades(n, p)
sm_data = {'n':n, 'p':p, 'tau':0.5, 'a':1, 'b':0.5, 'X':X}

# Initialise stan object
with open('student_grade_model.stan','r',newline='') as f:
    sm = stan.build(f.read(),sm_data,random_seed=1)

# Select the number of MCMC chains and iterations, then sample
chains, samples, burn = 4, 10000, 1000
fit=sm.sample(num_chains=chains, num_samples=samples, num_warmup=burn, save_warmup=False)

def plot_samples(fit,par,name,true_val=None):
    fig,axs=plt.subplots(2,2,figsize=(10,4),constrained_layout=True)
    fig.canvas.manager.set_window_title('Posterior for '+par)
    for i,j in [(i,j) for i in range(2) for j in range(2)]:
        axs[i,j].autoscale(enable=True, axis='x', tight=True)
    axs[0,0].set_title('Trace plot of log posterior density')
    axs[0,1].set_title('Trace plot of posterior samples of '+name)
    axs[1,0].set_title('Convergence of chain averages for '+name)
    axs[1,1].set_title('Approximate posterior density of '+name)
    par_mx=fit[par].reshape(samples,chains)
    lp_mx=fit['lp__'].reshape(samples,chains)
    for i in range(chains):
        x=i*samples+np.arange(samples)
        axs[0,0].plot(x,lp_mx[:,i])
        axs[0,1].plot(x,par_mx[:,i])
        axs[1,0].plot(x,np.cumsum(par_mx[:,i])/range(1,samples+1))
    axs[1,1].hist(par_mx.flatten(),200, density=True);
    if true_val is not None:
        axs[1,1].axvline(true_val, color='c', lw=2, linestyle='--')
    plt.show()
plot_samples(fit,'mu',r'$\mu$',true_val=mu)
```

After importing the necessary packages, the code first simulates student grade data for $n = 30$ students taking $p = 5$ tests. Second, the code loads in the Stan probability model from `student_grade_model.stan`. The third block determines that four separate *parallel* Hamiltonian MCMC chains are to be run, each requesting 10,000 samples after discarding the first 1000; the call to `stan.model.sample()` then obtains the posterior samples.

The final code block creates plots from the posterior samples. The top two cells show trace plots of the log posterior density of the sampled parameters and the values of the parameter μ from (6.1). The chains demonstrate stability and good mixing. The bottom row contains a diagnostic plot for the four chains showing the convergence of the sample mean for estimating the posterior expectation of μ, and finally a histogram of the sampled values of μ, pooled across the four chains, for estimating the marginal density $\pi(\mu)$. The true value of μ used to simulate the test scores is indicated with a dashed line; note the relatively small sample size (of students and test scores) means this true value is not yet well estimated by $\pi(\mu)$.

6.3 Other Software Libraries

6.3.1 PyMC

PyMC is a probabilistic programming package for Python which focuses on simplifying Bayesian inference, primarily through Markov chain Monte Carlo and variational methods. The fourth version (PyMC4) is being built upon a back end of the C++ and Python based symbolic mathematics library *TensorFlow*,[6] after the Python back end *Theano* used by previous versions was discontinued. PyMC offers similar functionality to Stan, also using the No-U-Turn Sampler (Hoffman and Gelman 2014) Hamiltonian Monte Carlo algorithm as the default inference tool for continuous parameters. Some users prefer PyMC to Stan for its native Python implementation, whilst others prefer Stan for its enhanced computation speed (through implementation in C++) and extensive documentation.

6.3.2 Edward

Edward is a Python library for probabilistic modelling and inference, named after the statistician George Edward Pelham Box who pioneered iterative approaches to statistical modelling. Edward was developed using TensorFlow as a back end. TensorFlow is designed for developing and training so-called *machine learning* models, and Edward builds upon this to offer modelling using neural networks (including popular *deep learning* techniques) but also supports graphical models (*cf.* Chap. 3) and Bayesian nonparametrics (*cf.* Chap. 9).

Bayesian inference can be performed using variational methods and Hamiltonian Monte Carlo, along with some other advanced techniques. Another strength of Edward is model criticism, using *posterior predictive checks* which assess how well data generated from the model under the posterior distribution agree with the realised data.

---

[6] https://www.tensorflow.org.

Chapter 7
Criticism and Model Choice

In subjective probability, there are no right or wrong systems of beliefs, provided they are coherently specified; I have my own individual measures of uncertainty concerning any quantities that I am unsure of, and it is fully admissible that these could be arbitrarily different from the probability beliefs held by others.

However, it was noted in the introductions to Chaps. 1 and 2 that mathematically specifying probability distributions which accurately represent systems of beliefs is a non-trivial exercise, and arguably always carries some degree of imprecision. The use of probability models, for example incorporating assumptions of exchangeability and making the choice of the prior measure Q in de Finetti's representation theorem (Sect. 2.2), provides tractable approximations of underlying beliefs which at least possess the necessary coherence properties for rational decision-making.

Furthermore, there is no philosophical requirement for subjective probability distributions to endure. They need only apply to the specific decision problem being addressed. Indeed, Bayes' theorem provides the coherent procedure for updating beliefs with new information with respect to a previously stated belief system. But for the next decision, there are other alternatives. In particular, I might want to review my previous decisions and the consequent outcomes, and call into question whether I should adopt a different perspective. Such considerations can be referred to as *model criticism* and *model selection*.

In this chapter, it is supposed that the decision-maker may be considering a range of modelling strategies for representing probabilistic beliefs about a random variable X for an uncertain outcome ω, where for sufficient generality $X : \Omega \to \mathbb{R}^n$ could represent a sequence of $n \geq 1$ real-valued observations. After observing the realised value $x = X(\omega)$, the decision-maker may want to re-evaluate which modelling strategy might have been most appropriate for capturing the true underlying dynamics which gave rise to x.

© The Author(s), under exclusive license to Springer Nature Switzerland AG 2021
N. Heard, *An Introduction to Bayesian Inference, Methods and Computation*,
https://doi.org/10.1007/978-3-030-82808-0_7

7.1 Model Uncertainty

Let $\mathcal{M}$ denote a set of models under consideration. Each proposed model $M \in \mathcal{M}$ corresponds to a probability distribution $\mathbb{P}(x \mid M)$ for the random outcome $X(\omega)$. Given an observed value of x, the quantity $\mathbb{P}(x \mid M)$ is known as the *evidence* for model M.

If M is a parametric model (4.1) with corresponding unknown parameter θ_M, then the model evidence $\mathbb{P}(x \mid M)$ can be regarded as a *marginal likelihood under model* M,

$$\mathbb{P}(x \mid M) = \int_{\Theta_M} p_M(x \mid \theta_M) \, p_M(\theta_M) \, d\theta_M, \qquad (7.1)$$

averaging the parametric likelihood $p_M(x \mid \theta_M)$ for model M with respect to the corresponding prior parameter density $p_M(\theta_M)$ for model M. Individual parametric models may therefore correspond to differences in one or more of the following:

- Underlying parameterisations, Θ_M;
- Likelihood model, $p_M(x \mid \theta_M)$;
- Prior distribution, $p_M(\theta_M)$.

Furthermore, it should be noted that not all rival models need be parametric, assume exchangeability, or presume any other structural similarities.

Two contrasting viewpoints can be adopted for handling Bayesian model uncertainty; the first allows all models in $\mathcal{M}$ to be considered simultaneously, known as *model averaging*, whilst the second, *model selection*, proposes a single chosen model from $\mathcal{M}$ for use in further analyses.

7.2 Model Averaging

To coherently consider all models in $\mathcal{M}$ simultaneously, the decision-maker must assert a subjective probability distribution Q over $\mathcal{M}$. Combining the probability distributions for the random variable X implied by the individual models, weighted according to these *prior* model probabilities, yields a marginal probability for X,

$$\mathbb{P}(x) = \int_{\mathcal{M}} \mathbb{P}(x \mid M) \, dQ(M). \qquad (7.2)$$

The expression (7.2) can be also be viewed as a marginal likelihood, averaging individual model marginal likelihoods (7.1) with respect to prior model uncertainty. In this sense, (7.2) is an example of using a mixture prior distribution (*cf.* Sect. 2.3.4). The averaging over a mixture of probability models (7.2) is known as *Bayesian model averaging*.

If the decision-maker is comfortable assigning probabilities across $\mathcal{M}$, the set of candidate models, and prepared to carry the extra computational burden of averaging across models to obtain marginal probability distributions (7.2), then this model-averaging approach is the correct method for managing model uncertainty under the Bayesian paradigm; the model uncertainty is simply one component of a mixture prior formulation.

7.3 Model Selection

Once the outcome variable x is observed, then if prior probabilities over $\mathcal{M}$ have been specified, the updated posterior model probabilities can be obtained via Bayes' theorem,

$$dQ(M \mid x) \propto dQ(M) \, \mathbb{P}(x \mid M).$$

In particular, if $\mathcal{M} = \{M_1, \ldots, M_k\}$ is a finite collection of k candidate models with prior probabilities $\mathbb{P}(M_1), \ldots, \mathbb{P}(M_k)$, then the posterior probability for the ith model can be expressed as

$$\mathbb{P}(M_i \mid x) = \frac{\mathbb{P}(M_i) \, \mathbb{P}(x \mid M_i)}{\sum_{i'=1}^{k} \mathbb{P}(M_{i'}) \, \mathbb{P}(x \mid M_{i'})}, \tag{7.3}$$

where the denominator is the model-averaged marginal likelihood (7.2).

7.3.1 Selecting From a Set of Models

If the decision problem is to determine which model was the underlying generative process which gave rise to x, then the decision-maker should proceed in the manner described in Chap. 1: specifying a utility or loss function which evaluates the consequences of estimating the model correctly or incorrectly and reporting the model which maximises expected utility with respect to the model posterior distribution (7.3).

Example 7.1 If choosing a model m from a finite set of models $\mathcal{M}$ using a zero-one utility function (*cf.* Exercise 1.8) with the following utility if the true model were M,

$$u(m, M) = \begin{cases} 1 & \text{if } m = M \\ 0 & \text{if } m \neq M \end{cases} \tag{7.4}$$

then the posterior expected utility of choosing model m is

$$\bar{u}(m) = \sum_{M \in \mathcal{M}} u(m, M) \, \mathbb{P}(M \mid x) = \mathbb{P}(m \mid x)$$

and the optimal Bayesian decision would be to report the posterior mode,

$$\arg\max_{M \in \mathcal{M}} \ \mathbb{P}(M \mid x).$$

7.3.2 Pairwise Comparisons: Bayes Factors

Suppose the decision-maker wishes to compare the relative suitability of two particular models, M_i and M_j; in this case, the comparison can be suitably encapsulated by the ratio of the posterior probabilities attributed to the two models.

Definition 7.1 (*Posterior odds ratio*) The *posterior odds ratio* of model M_i over model M_j is

$$\frac{\mathbb{P}(M_i \mid x)}{\mathbb{P}(M_j \mid x)} = \frac{\mathbb{P}(M_i)}{\mathbb{P}(M_j)} \times \frac{\mathbb{P}(x \mid M_i)}{\mathbb{P}(x \mid M_j)} \tag{7.5}$$

Remark 7.1 The first term on the right-hand side of (7.5) is known as the *prior odds ratio*, and the second term is known as the *Bayes Factor*.

Definition 7.2 (*Bayes factor*) The *Bayes factor* in favour of model M_i over model M_j is

$$B_{ij}(x) := \frac{\mathbb{P}(x \mid M_i)}{\mathbb{P}(x \mid M_j)} = \frac{\mathbb{P}(M_i \mid x)}{\mathbb{P}(M_j \mid x)} \bigg/ \frac{\mathbb{P}(M_i)}{\mathbb{P}(M_j)}.$$

Remark 7.2 The Bayes factor represents the evidence provided by the data x in favour of model M_i over M_j, measured by the multiplicative change observed in the odds ratio of the two models upon observing x.

If $B_{ij} > 1$, this suggests M_i has become more plausible relative to M_j after observing x, whereas $B_{ij} < 1$ suggests the opposite. Bayes factors are non-negative but have no upper bound, and although a larger Bayes factor presents stronger evidence in favour of M_i, there is no objective interpretation for any non-degenerate value. To provide interpretability, Jeffreys (1961) provided some subjective categorisations, which were later refined by Kass and Raftery (1995); the latter are shown in Table 7.1.

Table 7.1 Bayes factor interpretations according to Kass and Raftery (1995)

Bayes factor B_{ij}	Evidence in favour of M_i
1 to 3	Not worth more than a bare mention
3 to 20	Positive
20 to 150	Strong
>150	Very strong

7.3.2.1   Bayesian Hypothesis Testing

If one model M_0 is a special case of an alternative model M_1 (for example, M_0 assigns all probability mass to certain values of one or more free parameters in M_1), then the selection of a model is analogous to the frequentist statistical paradigm of hypothesis testing. The Bayes factor corresponds to the uniformly most powerful likelihood ratio test statistic.

Consider the zero-one utility function (7.4), for which the Bayes optimal decision is to report the most probable model. In hypothesis testing language, this implies the null model M_0 being rejected in favour of M_1 if and only if

$$\mathbb{P}(M_1 \mid x) > \mathbb{P}(M_0 \mid x) \iff B_{10}(x) > \frac{\mathbb{P}(M_0)}{\mathbb{P}(M_1)}. \tag{7.6}$$

The test procedure in (7.6) implies rejection of the null model M_0 in favour of M_1 if the Bayes factor $B_{10}(x)$ exceeds the prior ratio in favour of the null model. In this way, the prior ratio can be seen to determine the desired *significance level* of the test. A threshold value could be chosen by referring to the Bayes factor interpretations from Table 7.1.

Exercise 7.1 (*Bayes factors for Gaussian distributions*) Consider the following model for two exchangeable groups of random samples $\mathbf{x} = (x_1, \ldots, x_n)$, $\mathbf{y} = (y_1, \ldots, y_n)$:

$$
\begin{aligned}
x_i &\sim N(\theta_X, 1), \quad i = 1, \ldots, n, \\
y_i &\sim N(\theta_Y, 1), \quad i = 1, \ldots, n, \\
\theta_X, \theta_Y &\sim N(0, \sigma^2).
\end{aligned}
\tag{7.7}
$$

The samples $x_1, \ldots, x_n$ and $y_1, \ldots, y_n$ are all assumed to be conditionally independent given θ_X and θ_Y, and the model specification is completed by specifying the dependency between θ_X and θ_Y in one of two ways:

$$
\begin{aligned}
M_0 &: \theta_X = \theta_Y; \\
M_1 &: \theta_X \perp\!\!\!\perp \theta_Y.
\end{aligned}
\tag{7.8}
$$

(i) Derive an equation for the Bayes factor $B_{01}(\mathbf{x}, \mathbf{y})$ in favour of M_0 over M_1.
(ii) For fixed observed samples $\mathbf{x}$ and $\mathbf{y}$, show that $B_{01}(\mathbf{x}, \mathbf{y}) \to \infty$ as the assumed variance for the mean parameters θ_X and θ_Y, σ^2, tends to infinity. Comment.

Remark 7.3 The phenomenon mentioned in Exercise 7.1 Item ii is known as *Lindley's paradox*, named after the Bayesian decision theorist Dennis V. Lindley (1923–2013), and is further discussed in Proposition 8.3 of Chap. 8. For Bayesian hypothesis testing, there is no useful concept of a totally uninformative prior for model selection. If beliefs about unknown parameters are made arbitrarily vague, then the simpler model will *always* be preferred, regardless of the data.

7.3.3   *Bayesian Information Criterion*

One issue with using posterior probabilities and Bayes factors for choosing amongst models is that these quantities rely upon calculation of the marginal likelihoods of observed data for each model. It was noted in Sect. 4.1 that the marginal likelihood will not be analytically calculable for most models; and although Sect. 5.2.4.1 proposed numerical importance sampling methods for estimating marginal likelihoods, reliable low variance estimates may not be available.

Suppose $x = (x_1, \ldots, x_n)$. When the number of samples n is large, Schwarz (1978) showed that for exponential family (*cf.* Sect. 4.3) models with a k-dimensional parameter θ,

$$\log p(x) \approx \log p(x \mid \hat{\theta}) - \frac{k}{2} \log n,$$

where $\hat{\theta}$ is the maximum likelihood estimate of θ maximising $p(x \mid \theta)$.

On this basis, a popular method for comparing rival models (even outside of exponential families) is the so-called Bayesian information criterion.

Definition 7.3 The *Bayesian information criterion* for model selection is defined to be

$$\text{BIC} := -2 \log p(x \mid \hat{\theta}) + k \log n \tag{7.9}$$

where k is the dimension of θ and $\hat{\theta}$ maximises $p(x \mid \theta)$. Low BIC values correspond to good model fit.

Remark 7.4 For a given likelihood model, the BIC (7.9) is twice the negative logarithm of an asymptotic approximation of a corresponding Bayesian marginal likelihood for n samples as $n \to \infty$; this asymptotic marginal likelihood does not depend on the choice of prior $p(\theta)$, besides requiring appropriate support for the maximum likelihood estimate. The BIC is therefore only suitable for comparing different formulations of the likelihood component of a parametric model, and not for comparing prior distributions.

Proposition 7.1 (BIC approximated Bayes factors) *If BIC_i and BIC_j denote the Bayesian information criterion for two models M_i and M_j, an approximate Bayes factor in favour of model i over model j is*

$$B_{ij} \approx \exp\left\{ -\frac{1}{2}(\text{BIC}_i - \text{BIC}_j) \right\}.$$

Exercise 7.2 (*BIC for Gaussian distributions*) Consider the sampling model (7.7) for two groups of random samples $\mathbf{x} = (x_1, \ldots, x_n)$, $\mathbf{y} = (y_1, \ldots, y_n)$ presented in Exercise 7.1, and the two alternative models M_0 and M_1 from (7.8) for the respective mean parameters θ_X and θ_Y.

(i) Derive equations for the Bayesian information criterion values BIC_0 and BIC_1 for the two models M_0 over M_1.

(ii) Use these BIC values to give an approximate expression for the corresponding Bayes factor B_{01}.

7.4 Posterior Predictive Checking

Posterior probabilities and Bayes factors are useful for assessing the relative merits of rival models. However, it can also be desirable to assess the quality of a single model in absolute terms, without reference to any proposed alternatives which may not yet have been identified. Posterior predictive checking (PPC) methods aim to quantify how well a proposed model structure fits the observed data, using the following logic: if the model is a good approximation to the generating mechanism for the observed data, then the posterior distribution of the model parameters should assign high probability to parameter values which in turn would generate further data similar to the observed data with high probability if the sampling process was repeated.

Consider a single parametric model with model parameter $\theta \in \Theta$, prior density $p(\theta)$ and likelihood density $p(x \mid \theta)$ for the observed data $x \in \mathcal{X}$. Let $\pi(\theta)$ be the corresponding posterior density (4.3).

Definition 7.4 (*Posterior predictive distribution*) The *posterior predictive distribution* is the marginal distribution of a second draw $x_{\text{rep}} \in \mathcal{X}$ from the likelihood model with the same (unknown) parameter, implying a density

$$\pi(x_{\text{rep}}) := \int_{\Theta} p(x_{\text{rep}} \mid \theta) \, \pi(\theta) \, d\theta$$

$$\propto \int_{\Theta} p(x_{\text{rep}} \mid \theta) \, p(x \mid \theta) \, p(\theta) \, d\theta. \tag{7.10}$$

Using techniques from frequentist statistical hypothesis testing, posterior predictive checking is concerned with establishing whether the observed data x could be regarded as being somehow *extreme* with respect to the posterior predictive density (7.10).

7.4.1 Posterior Predictive p-Values

For full generality, let $T(x, \theta)$ be a test statistic for measuring discrepancy between a data-generating parameter θ and observing data x.

Definition 7.5 (*Posterior predictive p-value*) A *posterior predictive p-value* for $T(x, \theta)$ is the upper tail probability

$$p := \int_{\Theta} \int_{\mathcal{X}} \mathbb{1}_{[T(x,\theta),\infty)}\{T(x_{\text{rep}}, \theta)\} \, \pi(\theta) \, p(x_{\text{rep}} \mid \theta) \, dx_{\text{rep}} \, d\theta. \tag{7.11}$$

Remark 7.5 If the test statistic is simply a function of the data, $T(x, \theta) \equiv T(x)$, then (7.11) simplifies to

$$p = \int_{\mathscr{X}} \mathbb{1}_{[T(x), \infty)} \{T(x_{\text{rep}})\} \, \pi(x_{\text{rep}}) \, dx_{\text{rep}},$$

which is the familiar one-sided p-value for an observed statistic $T(x)$, calculated with respect to the posterior predictive distribution (7.10).

Remark 7.6 More generally, the posterior predictive p-value (7.11) measures how a joint sample of parameter and new data from the posterior would compare with sampling a parameter from the posterior and pairing this with the observed data.

7.4.2 Monte Carlo Estimation

Given (possibly approximate) samples $\theta^{(1)}, \ldots, \theta^{(m)}$ obtained from the posterior density π, a Monte Carlo estimate (*cf.* Sect. 5.2) of the posterior predictive p-value (7.11) can be obtained relatively easily provided it is also possible to sample from the likelihood distribution $p(x \mid \theta)$: For each parameter value $\theta^{(i)}$ sampled from the posterior density π, randomly draw new data $x_{\text{rep}}^{(i)}$ from the generative likelihood model with that parameter,

$$x_{\text{rep}}^{(i)} \sim p(x_{\text{rep}}^{(i)} \mid \theta^{(i)}); \tag{7.12}$$

then the Monte Carlo estimated posterior predictive p-value is

$$\hat{p} := \frac{1}{m} \sum_{i=1}^{m} \mathbb{1}_{[T(x, \theta^{(i)}), \infty)} \{T(x_{\text{rep}}^{(i)}, \theta^{(i)})\}. \tag{7.13}$$

7.4.3 PPC with Stan

When fitting Bayesian models numerically in Stan (*cf.* Sect. 6.2), it is relatively simple to carry out posterior predictive checking using a `generated quantities{}` code block. This will be illustrated for the student grades example in Sect. 6.2, by considering two possible test statistics: The first test statistic uses the negative log likelihood as a measure of discrepancy,

$$T(x, \theta) = -\log \, p(x \mid \theta). \tag{7.14}$$

The second statistic does not depend on the model parameters, simply obtaining the average score for each student,

$$\bar{x}_i = \frac{1}{p} \sum_{j=1}^{p} x_{ij},$$

and then calculating the variance of these scores

$$T(x) = \frac{n \sum_{i=1}^{n} \bar{x}_i^2 - (\sum_{i=1}^{n} \bar{x}_i)^2}{n(n-1)}. \tag{7.15}$$

The following Stan programming code (`student_grade_model_ppc.stan`) extends the example from Sect. 6.2 with the inclusion of a `generated quantities{}` code block to facilitate posterior predictive checks using the test statistics (7.14) and (7.15).

```
1   // student_grade_model_ppc.stan
2
3   data {
4       int<lower=0> n; // number of students
5       int<lower=0> p; // number of tests
6       int<lower=0, upper=100> X[n, p]; // student test grades
7       real<lower=0> tau;
8       real<lower=0> a;
9       real<lower=0> b;
10  }
11  parameters {
12      real z[n];
13      real mu;
14      real<lower=0> sigma_sq;
15  }
16  transformed parameters {
17      real<lower=0, upper=1> theta[n];
18      real sigma;
19      theta = inv_logit(z);
20      sigma = sqrt(sigma_sq);
21  }
22  model {
23      sigma_sq ~ inv_gamma(a,b);
24      mu ~ normal(0, sigma * tau);
25      z ~ normal(mu, sigma);
26      for (i in 1:n)
27          X[i] ~ binomial(100,theta[i]);
28  }
29  generated quantities{
30      int<lower=0, upper=100> X_rep[n, p];
31      real log_lhd = 0;
32      real log_lhd_rep = 0;
33      real ppp;
34      for (i in 1:n){
35          for (j in 1:p){
36              log_lhd += binomial_lpmf(X[i][j] | 100,theta[i]);
37              X_rep[i][j] = binomial_rng(100,theta[i]);
38              log_lhd_rep += binomial_lpmf(X_rep[i][j] | 100,theta[i]);
39          }
40      }
41      ppp = log_lhd >= log_lhd_rep ? 1 : 0;
42  }
```

Line number 37 of `student_grade_model_ppc.stan` generates a replicate data matrix X_{rep} from the binomial model with the current sampled parameter vector θ; this is required for both test statistics (7.14) and (7.15). To calculate the test statistic (7.14) within Stan, lines 36 and 38 calculate the likelihood function for the original and replicated data matrices, respectively, and these are compared on line 41, yielding an indicator to contribute towards the estimated posterior predictive p-value (7.13).

The following PyStan code (`student_grade_inference_stan_ppc.py`) uses the Stan model code from above to fit the model and perform posterior predictive checking.

```python
#! /usr/bin/env python
## student_grade_inference_stan_ppc.py

import stan
import numpy as np
import matplotlib.pyplot as plt

# Simulate data
from student_grade_simulation import sample_student_grades
n, p = 30, 5
X, mu, sigma = sample_student_grades(n, p)
sm_data = {'n':n, 'p':p, 'tau':0.5, 'a':1, 'b':0.5, 'X':X}

# Initialise stan object
with open('student_grade_model_ppc.stan','r',newline='') as f:
    sm = stan.build(f.read(),sm_data,random_seed=1)

# Select the number of MCMC chains and iterations, then sample
chains, samples, burn = 4, 10000, 1000
fit=sm.sample(num_chains=chains, num_samples=samples, num_warmup=burn, save_warmup=False)

def T(x): #Variance of student average scores
    return(np.var(np.mean(x,axis=1)))

t_obs = T(X) #Value of test statistic for observed data
x_rep = fit['X_rep'].reshape(n,p,samples,chains)
t_rep = [[T(x_rep[:,:,i,j]) for i in range(samples)] for j in range(chains)]

# Plot posterior predictive distributions of T from each chain
def posterior_predictive_plots(t_rep,true_val):
    nc = np.matrix(t_rep).shape[0]
    fig,axs=plt.subplots(1,nc,figsize=(10,3),constrained_layout=True)
    fig.canvas.manager.set_window_title('Posterior predictive')
    for j in range(nc):
        axs[j].autoscale(enable=True, axis='x', tight=True)
        axs[j].set_title('Chain '+str(j+1))
        axs[j].hist(np.array(t_rep[j]),200, density=True)
        axs[j].axvline(true_val, color='c', lw=2, linestyle='--')
    plt.show()

posterior_predictive_plots(t_rep,t_obs)

# Calculate and print posterior predictive p-values for T
print("Posterior predictive p-values from variance of means:")
print([np.mean(t_obs > t_rep[j]) for j in range(chains)])

# Print posterior predictive p-values for lhd calculated in Stan
print("Posterior predictive p-values from likelihood:")
print(np.mean(fit['ppp'].reshape(samples,chains),axis=0))
```

Line numbers 22–23 of `student_grade_inference_stan_ppc.py` define the student variance test statistic (7.15); this is evaluated for the observed data matrix at line 25, and for each of the data matrix replicates sampled from the posterior predictive distribution at line 27. At line 41, the estimated posterior predictive distribution of the variance for a new student cohort is plotted for each MCMC chain, and compared with the observed value. Finally, lines 45 and 49 print the estimated posterior predictive p-values for each of the two test statistics, obtained from each of the four MCMC chains.

The following outputs from the code were obtained:

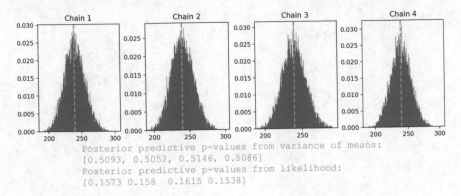

Posterior predictive p-values from variance of means:
[0.5093, 0.5052, 0.5146, 0.5086]
Posterior predictive p-values from likelihood:
[0.1573 0.158 0.1615 0.1538]

The *p*-values suggest no statistical significance for either test statistic, suggesting a good model fit; indeed, the data have been generated from the assumed probability model.

Chapter 8
Linear Models

Infinite exchangeability of a sequence of random variables, here denoted $y_1, y_2, \ldots$, is a useful simplifying assumption for illustrating many of the fundamental ideas presented in the preceding chapters. However, in many practical situations, this would be too limiting as a modelling assumption; often there will be additional available information x_i pertaining to each random quantity y_i which affects probabilistic beliefs about the value which y_i is likely to take.

In the language of statistical *regression* modelling, the random variables of interest $y_1, y_2, \ldots$ are referred to as *response* variables; they are believed to have a statistical dependence on the corresponding element of the sequence of so-called *covariates* or *predictors* $x_1, x_2, \ldots$ which have either been determined or observed. Regression modelling is concerned with building statistical models for the conditional distribution of each y_i given x_i, primarily through specifying the mean value for y_i having some functional relationship to x_i (referred to as the *regression function*).

The simplest functional relationship is the linear model. With assumed Gaussian *errors* in the response variable, the elegant least squares estimation equations from non-Bayesian statistical linear models extend naturally to the Bayesian case. Despite the apparent rigidity of a linearity assumption, consideration of different transformations of either the covariates or the response variable can provide a surprisingly flexible modelling framework.

8.1 Parametric Regression

Let $y = (y_1, \ldots, y_n)$ be an n-vector of real-valued response variables. For each response variable $y_i \in \mathbb{R}$, suppose there is a corresponding p-vector of covariates $x_i = (x_{i1}, \ldots, x_{ip}) \in \mathbb{R}^p$, $p \geq 1$, which are thought to provide information about

---

The original version of this chapter has been revised due to typographic errors. The corrections to this chapter can be found at https://doi.org/10.1007/978-3-030-82808-0_12

Fig. 8.1 A belief network
representation of regression
exchangeability for
responses $y_1, \ldots, y_n$ given
covariates $x_1, \ldots, x_n$

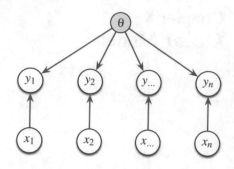

the probability distribution of y_i. Let $X = (x_{ij})$ be an $n \times p$ matrix with ith row x_i corresponding to the ith response.

   In parametric regression modelling, it is common to assume a relaxation of exchangeability called *regression exchangeability*.

Definition 8.1 (*Regression exchangeability*) Regression exchangeability assumes that the joint density for y conditional on X has a representation

$$p(y \mid X) = \int_{\Theta} \prod_{i=1}^{n} p(y_i \mid \theta, x_i) \, dQ(\theta) \qquad (8.1)$$

for a prior distribution Q on some parameter space Θ.

   The regression function relating the response to the covariates is specified through the likelihood density $p(y_i \mid \theta, x_i)$ in (8.1). Figure 8.1 shows a belief network representation of regression exchangeability.

8.2   Bayes Linear Model

The linear model is a special case of (8.1), where the parameter is a pair $\theta = (\beta, \sigma)$, with $\beta \in \mathbb{R}^p$ and $\sigma > 0$, and the likelihood density $p(y_i \mid \theta, x_i)$ is specified by

$$y_i \mid \theta, x_i \sim \text{Normal}(x_i \cdot \beta, \sigma^2). \qquad (8.2)$$

The parameter σ is the standard deviation of each response variable, and β is a p-vector of regression coefficients such that

$$\mathbb{E}(y_i \mid x_i, \beta) = x_i \cdot \beta. \qquad (8.3)$$

The model (8.2) is often written in an equivalent regression form:

$$y_i = f(x_i) + \epsilon_i,$$
$$\epsilon_i \sim \text{Normal}(0, \sigma^2),$$

where

$$f(x) = x \cdot \beta \tag{8.4}$$

is the regression function and $\epsilon = (\epsilon_1, \ldots, \epsilon_n)$ are referred to as independent *error* variables.

Finally, a third way of expressing the same model, which can be particularly convenient for mathematical manipulation, is in matrix form

$$y \sim \mathrm{Normal}_n(X\beta, \sigma^2 I_n), \tag{8.5}$$

where the conditional independence of the response variables in (8.1) is represented by the diagonal covariance matrix in (8.5).

For a Bayesian parametric regression model (8.1), the specification of the linear model is completed by a prior distribution for the parameters (β, σ). Two choices of prior are commonly considered, which are presented in the next two subsections.

8.2.1 Conjugate Prior

Since (8.5) is an exponential family distribution (*cf.* Sect. 4.3), it follows from Proposition 4.2 that there is a conjugate prior for (β, σ). This takes a canonical form

$$\beta \mid \sigma \sim \mathrm{Normal}_p(0, \sigma^2 V),$$
$$\sigma^{-2} \sim \mathrm{Gamma}(a, b), \tag{8.6}$$

where V is a symmetric, positive semidefinite $p \times p$ covariance matrix and $a, b > 0$.

Exercise 8.1 (*Marginal density for regression coefficients*) Suppose the conjugate prior distribution (8.6) for the normal linear model.

(i) Show that, marginally,

$$p(\beta) = \frac{\Gamma(a + \frac{p}{2})}{(2\pi b)^{\frac{p}{2}} |V|^{\frac{1}{2}} \Gamma(a)} \left(1 + \frac{\beta^\mathsf{T} V^{-1} \beta}{2b} \right)^{-(a + \frac{p}{2})},$$

corresponding to the density of the multivariate t-distribution, $\beta \sim t_{2a}(0, \frac{b}{a} V)$.

(ii) If $V = I_p$, show that the prior density function for β depends only on the Euclidean norm $\|\beta\| = \sqrt{\beta \cdot \beta}$ of the regression coefficients and that

$$\sqrt{\frac{2a + p - 1}{2b}} \, \|\beta\| \sim |t_{2a}|, \tag{8.7}$$

known as a *half-Student's t-distribution* with $2a$ degrees of freedom.

Typically the regression coefficients will be assumed to be independent and identically distributed, which corresponds to assuming

$$V = \lambda^{-1} I_p \tag{8.8}$$

for a scalar *precision* parameter $\lambda > 0$. This implies a joint prior probability density function

$$p(\beta, \sigma^{-2}) = \frac{b^a \, \lambda^{\frac{p}{2}} \, \exp\left\{-\sigma^{-2} \, (2b + \lambda\beta^{\mathsf{T}} \cdot \beta) \, /2\right\}}{(2\pi)^{\frac{p}{2}} \, \Gamma(a) \, \sigma^{2(a-1)+p}}. \tag{8.9}$$

Proposition 8.1 *For the linear model (8.5) with conjugate prior (8.6), the posterior distribution for (β, σ) after observing responses $y = (y_1, \ldots, y_n)$ corresponds to*

$$\beta \mid \sigma, X, y \sim \text{Normal}_p(m_n, \sigma^2 V_n),$$
$$\sigma^{-2} \mid X, y \sim \text{Gamma}(a_n, b_n),$$

where

$$V_n = (V^{-1} + X^{\mathsf{T}}X)^{-1}, \qquad m_n = V_n \, X^{\mathsf{T}}y,$$
$$a_n = a + \frac{n}{2}, \qquad\qquad b_n = b + \frac{1}{2}(y^{\mathsf{T}}y - y^{\mathsf{T}}Xm_n). \tag{8.10}$$

Proof

$$p(\beta \mid \sigma, X, y) \propto p(\beta \mid \sigma) \, p(y \mid \beta, \sigma, X)$$

$$\propto \exp\left\{-\frac{1}{2\sigma^2}\beta^{\mathsf{T}}V^{-1}\beta - \frac{1}{2\sigma^2}(y - X\beta)^{\mathsf{T}}(y - X\beta)\right\}$$

$$\propto \exp\left[-\frac{1}{2\sigma^2}\{\beta^{\mathsf{T}}(V^{-1} + X^{\mathsf{T}}X)\beta - y^{\mathsf{T}}X\beta - \beta^{\mathsf{T}}X^{\mathsf{T}}y\}\right]$$

$$= \exp\left[-\frac{1}{2\sigma^2}\{\beta^{\mathsf{T}}V_n^{-1}\beta - m_n^{\mathsf{T}}V_n^{-1}\beta - \beta^{\mathsf{T}}V_n^{-1}m_n\}\right],$$

according to (8.10) since $V_n^{-1}m_n = X^{\mathsf{T}}y$. Completing the square,

$$p(\beta \mid \sigma, X, y) \propto \exp\left\{-\frac{1}{2\sigma^2}(\beta - m_n)^{\mathsf{T}}V_n^{-1}(\beta - m_n)\right\}$$
$$\implies \beta \mid \sigma, y \sim \text{Normal}_p(m_n, \sigma^2 V_n).$$

It remains to derive the posterior distribution of σ. The regression coefficients β can be marginalised,

$$\beta \mid \sigma \sim \text{Normal}_p(0, \sigma^2 V)$$
$$\implies X\beta \mid \sigma, X \sim \text{Normal}_n(0, \sigma^2 X V X^\mathsf{T})$$
$$\implies y \mid \sigma, X \sim \text{Normal}_n(0, \sigma^2(X V X^\mathsf{T} + I_n)), \tag{8.11}$$

where the last step follows from standard rules for summing Gaussian random variables. Then by the matrix inversion lemma,

$$(X V X^\mathsf{T} + I_n)^{-1} = I_n - X(V^{-1} + X^\mathsf{T} X)^{-1} X^\mathsf{T} = I_n - X V_n X^\mathsf{T}$$
$$\implies y \mid \sigma, X \sim \text{Normal}_n(0, \sigma^2(I_n - X V_n X^\mathsf{T})^{-1}).$$

By Bayes' theorem,

$$p(\sigma^{-2} \mid X, y) \propto p(\sigma^{-2}) \, p(y \mid \sigma, X)$$

$$\propto \sigma^{-2(a-1)} \exp(-b\sigma^{-2}) \, \sigma^{-n} \exp\left\{-\frac{\sigma^{-2}}{2} y^\mathsf{T}(I_n - X V_n X^\mathsf{T}) y\right\}$$

$$= \sigma^{-2(a+\frac{n}{2}-1)} \exp\left[\sigma^{-2}\left\{\frac{1}{2} y^\mathsf{T}(I_n - X V_n X^\mathsf{T}) y\right\}\right]$$

$$\implies \sigma^{-2} \mid X, y \sim \text{Gamma}(a_n, b_n).$$

Exercise 8.2 (*Linear model matrix inverse*) The matrix inversion lemma states that for an $n \times n$ matrix A, a $k \times k$ matrix V and $n \times k$ matrices U, W,

$$(A + U V W^\mathsf{T})^{-1} = A^{-1} - A^{-1} U (V^{-1} + W^\mathsf{T} A^{-1} U)^{-1} A^{-1}.$$

Using this result, show that $(X V X^\mathsf{T} + I_n)^{-1} = I_n - X V_n X^\mathsf{T}$ where $V_n = (V^{-1} + X^\mathsf{T} X)^{-1}$.

From Proposition 4.1, it follows that the Bayes linear model with conjugate prior has a closed-form marginal likelihood.

Proposition 8.2 *Suppose the Bayes linear model* (8.5) *with* $y \in \mathbb{R}^n$, $X \in \mathbb{R}^{n \times p}$ *and conjugate prior* (8.6). *The marginal likelihood for* $y \mid X$ *is*

$$p(y \mid X) = \frac{\Gamma(a_n) \, |V_n|^{\frac{1}{2}} \, b^a}{(2\pi)^{\frac{n}{2}} \, \Gamma(a) \, |V|^{\frac{1}{2}} \, b_n^{a_n}}. \tag{8.12}$$

Equivalently,

$$y \mid X \sim \text{St}_n(2a, 0, b(X V X^\mathsf{T} + I_n)/a),$$

where $\text{St}_n(v, \mu, \Sigma)$ *is an* n-*dimensional Student's* t-*distribution with* v *degrees of freedom, mean* μ *and covariance* Σ.

Proof From the proof of Proposition 8.1,

$$y \mid \sigma, X \sim \text{Normal}_n (0, \sigma^2 (I_n - X V_n X^\mathsf{T})^{-1})$$

$$\implies p(y \mid \sigma, X) = \frac{\exp\{-\frac{1}{2\sigma^2} y^\mathsf{T} y + \frac{1}{2\sigma^2} y^\mathsf{T} X (V^{-1} + X^\mathsf{T} X)^{-1} X^\mathsf{T} y\}}{(2\pi)^{\frac{n}{2}} \sigma^n |V|^{\frac{1}{2}} |V^{-1} + X^\mathsf{T} X|^{\frac{1}{2}}}. \qquad (8.13)$$

The denominator uses the identity $|X V X^\mathsf{T} + I_n| = |V||V^{-1} + X^\mathsf{T} X|$, which follows from the matrix determinant lemma (Proposition 8.3).

Marginalising (8.13) over the inverse-gamma prior for σ^2,

$$p(y \mid X) = \frac{b^a}{\Gamma(a)} \int_0^\infty \sigma^{-2(a-1)} \exp\{-b\sigma^{-2}\} \, p(y \mid \sigma, X) \, d\sigma^{-2}$$

$$= \frac{\Gamma(a + n/2) |V_n|^{\frac{1}{2}} b^a}{(2\pi)^{\frac{n}{2}} \Gamma(a) |V|^{\frac{1}{2}} (b + \frac{1}{2} y^\mathsf{T} y - \frac{1}{2} y^\mathsf{T} X m_n)^{a + \frac{n}{2}}},$$

by comparison of the integrand with the $\text{Gamma}(a + n/2, b + \frac{1}{2} y^\mathsf{T} y - \frac{1}{2} y^\mathsf{T} X m_n)$ density function.

Exercise 8.3 (*Linear model matrix determinant*) The matrix determinant lemma states that for an $n \times n$ matrix A, a $k \times k$ matrix V and $n \times k$ matrices U, W, $|A + U V W^\mathsf{T}| = |V^{-1} + W^\mathsf{T} U||V||A|$. Using this result, show that $|X V X^\mathsf{T} + I_n| = |V||V^{-1} + X^\mathsf{T} X|$.

Proposition 8.3 (Lindley's paradox) *For the linear model under the conjugate prior* (8.6) *and assuming* (8.8), *as* $\lambda \to 0$ *the marginal likelihood* (8.12) $p(y \mid X) \to 0$.

Proof As $\lambda \to 0$, $|V| \to \infty$ whilst $|V_n| \to 1/|X^\mathsf{T} X|$. Hence $p(y \mid X) \to 0$.

Remark 8.1 Lindley's paradox in Proposition 8.3 (*cf.* Proposition 7.1) states that making prior beliefs increasingly diffuse will eventually lead to diminishingly small predictive probability density for any possible observation y. Consequently, when comparing against any fixed alternative model, the Bayes factor in favour of the alternative model will become arbitrarily large.

🖳 **Exercise 8.4** (*Linear model code*) Write computer code (using a language such as Python) to calculate the marginal likelihood under the linear model. For a matrix of covariates X and a vector of responses y, write a single function which returns both the marginal likelihood and the posterior mean for the regression coefficients.

Exercise 8.5 (*Orthogonal covariate matrix marginal likelihood*) Suppose the columns of the matrix X are orthonormal. Then under model (8.9) where the regression coefficients are assumed to be independent, derive a simplified expression for the linear model marginal likelihood $p(y \mid X)$. Comment on why this expression should be easier to evaluate than the general expression (8.12).

Exercise 8.6 (*Zellner's g-prior*) Suppose the $n \times p$ covariate matrix X has rank p, with $n > p$, and the matrix V in (8.6) satisfies $V = g \cdot (X^\mathsf{T} X)^{-1}$ for some constant $g > 0$; this formulation is known as Zellner's g-prior (Zellner 1986). Derive a simplified expression for the linear model marginal likelihood $p(y \mid X)$ under this prior distribution.

8.2.2 Reference Prior

A commonly used alternative prior distribution is the uninformative *reference* prior (*cf.* Sect. 2.3.2),

$$p(\beta, \sigma^2) \propto \frac{1}{\sigma^2}, \tag{8.14}$$

corresponding to "uniform" prior beliefs for $\log \sigma^2$ and each component of the coefficient vector β.

Remark 8.2 The prior density (8.14) is said to be *improper* since it does not have a finite integral over the parameter space. It can therefore only be meaningfully considered as the limiting argument of a sequence of increasingly diffuse, proper prior densities.

The reference prior can be viewed as a limiting case of the conjugate prior (8.9) as the hyperparameters $a, b, \lambda \to 0$. Consequently, the posterior distribution result from Sect. 8.2.1 carries across as follows.

Proposition 8.4 *For the linear model* (8.5) *with reference prior* (8.14), *if* $n > p$ *and* X *has rank* p, *then the posterior distribution for* (β, σ) *is*

$$\beta \mid \sigma, X, y \sim \mathrm{Normal}_p((X^\mathsf{T} X)^{-1} X^\mathsf{T} y, \; \sigma^2 (X^\mathsf{T} X)^{-1}),$$
$$\sigma^{-2} \mid X, y \sim \mathrm{Gamma}(a + n/2, \; b + (y^\mathsf{T} y - y^\mathsf{T} X (X^\mathsf{T} X)^{-1} X^\mathsf{T} y)/2). \tag{8.15}$$

Remark 8.3 The reference posterior (8.15) is only proper when $n > p$ and the rank of X is equal to p, so that X has full rank.

Remark 8.4 As a direct consequence of Lindley's paradox in Proposition 8.3, the marginal likelihood is not well defined under the improper reference prior (8.14); the corresponding equation takes value 0 for all values of y. Consequently, reference priors cannot be used when performing model choice using Bayes factors (*cf.* Sect. 7.3.2).

Remark 8.5 Bayesian inference for the linear model under the reference prior corresponds to the standard estimation procedures from classical statistics. For example, the posterior mean for β in (8.15) is the usual least squares or maximum likelihood estimate.

8.3   Generalisation of the Linear Model

Aside from the reference prior analysis in Sect. 8.2.2, the theory of the Bayes linear model with conjugate prior required no assumptions about the nature of the covariates $x_i = (x_{i1}, \ldots, x_{ip}) \in \mathbb{R}^p$ which make up the rows of the matrix X. This observation allows the following abstraction of the so-called *design matrix* X, which provides a valuable generalisation in the use of the linear model.

8.3.1   General Basis Functions

Most generally, suppose that for each response variable y_i there is an observed p'-vector of related measurements $z_i \in \mathbb{R}^{p'}$, for $p' \geq 1$. Setting $p = p'$ and $x_i = z_i$ returns the standard linear model (8.5), which is linear in both β and the measurements z_i.

More generally, suppose a list of p functions $\psi = (\psi_1, \ldots, \psi_p)$,

$$\psi_j : \mathbb{R}^{p'} \to \mathbb{R}, \quad j = 1, \ldots, p,$$

such that each function specifies a covariate for the linear model, leading to the p-vector covariate

$$x_i = \psi(z_i) = (\psi_1(z_i), \ldots, \psi_p(z_i)).$$

The functions ψ are referred to as *basis functions*, since the regression function (8.4) is constructed by linear combinations of the components of ψ,

$$f(x) = \sum_{j=1}^{p} \beta_j \psi_j(z).$$

Each basis function ψ_j corresponds to a column of the resulting covariate matrix X. Given the flexibility available in the choice of the columns of X, the matrix X is often referred to as the *design matrix* for a regression model.

Remark 8.6 The linear model (8.5) with design matrix X specified by

$$X_{ij} = \psi_j(z_i)$$

is still a linear model with respect to the regression coefficients β, and so all of the preceding theory for conjugate posterior distributions and closed-form marginal likelihoods still applies.

8.3.1.1 Polynomial Regression

Suppose a single observable measurement $z \in \mathbb{R}$ and basis functions

$$\psi_j(z) = z^{j-1}, \quad j = 1, \ldots, p,$$

implying a covariate vector

$$x = (1, z, \ldots, z^{p-1}).$$

This construction implies a degree $p - 1$ polynomial regression function,

$$f(x) = \sum_{j=1}^{p} \beta_j \, z^{j-1},$$

as a special case of the Bayes linear model.

8.3.1.2 Linear Spline Regression

Let the notation $(\cdot)_+$ denote the positive part of a real number,

$$(t)_+ = \max\{0, t\}.$$

Again suppose a single observable measurement $z \in \mathbb{R}$ and now consider the basis functions

$$\psi_j(z) = (z - \tau_j)_+, \quad j = 1, \ldots, p, \tag{8.16}$$

for a sequence of p real values $\tau_1 < \tau_2 < \ldots < \tau_p$, referred to as *knot points*. Basis functions of the type (8.16) are known as linear splines, since $\psi_j(z)$ is zero up until the value τ_j, and a linear function of $z - \tau_j$ thereafter.

Taking a linear combination of linear spline basis functions gives a regression function which is piecewise linear, with changes in gradient occurring at each of the knot points but no discontinuities. Spline regression models are explored in more detail in Sect. 10.3.

8.4 Generalised Linear Models

The Bayes linear model presented in Sects. 8.2 and 8.3 is mathematically very convenient, but is only suitable for cases where the response variables can be assumed to be normally distributed and where the linear regression function $x_i \cdot \beta$ corresponds directly to the expected value of the response variable y_i (8.3).

Generalised linear models extend the linear model to other exponential family distributions for the response variable through the introduction of an invertible function g called the *link* function, such that

$$g\{\mathbb{E}(y_i \mid x_i, \beta)\} = x_i \cdot \beta,$$

or, equivalently,

$$\mathbb{E}(y_i \mid x_i, \beta) = g^{-1}(x_i \cdot \beta). \tag{8.17}$$

Remark 8.7 The link function generalises the standard linear regression expectation (8.3), which is clearly a special case of (8.17) with the identity link function.

Remark 8.8 One advantage of using a link function is to guarantee the expected value of the response (8.17) lies in the correct domain, without requiring any constraints on the possible values which the covariates x_i or the regression coefficients might take.

Two examples of generalised linear models are now briefly presented, where the response variable is either a non-negative integer count or a binary indicator. In both cases, a zero-mean normal distribution prior (8.6) with $V = I_p$ is assumed for the regression coefficients, which was shown in (8.7) to imply a t-distribution on the Euclidean norm of the coefficients.

8.4.1 Poisson Regression

Suppose each response $y_i \in \mathbb{N} = \{0, 1, 2, \ldots\}$ is a non-negative integer count. Further assume each count follows a Poisson distribution, with an expected value which is believed to be linearly dependent on $p \geq 1$ covariates $x_i \in \mathbb{R}^p$ through the link function $\log(\cdot)$. These assumptions imply

$$\log \mathbb{E}(y_i \mid x_i, \beta) = x_i \cdot \beta,$$

for some $\beta \in \mathbb{R}^p$ and

$$y_i \mid x_i, \beta \sim \text{Poisson}(\exp(x_i \cdot \beta))$$

$$\implies p(y_i \mid x_i, \beta) = \frac{\exp\{y_i \, (x_i \cdot \beta) - \exp(x_i \cdot \beta)\}}{y_i!}. \tag{8.18}$$

8.4.1.1 Stan Implementation

The Stan implementation of Poisson regression extends the model (8.18) slightly by assuming the presence of a variable intercept term in the linear model,

$$y_i \mid x_i, \alpha, \beta \sim \text{Poisson}(\exp(\alpha_i + x_i \cdot \beta)),$$

for $\alpha = (\alpha_1, \ldots, \alpha_n) \in \mathbb{R}^n$. However, these parameters can be fixed at zero to recover the standard representation (8.18).

Remark 8.9 In some non-Bayesian statistical texts, the intercept parameters α_i would be referred to as *random effects*, since they differ between individual response variables, whilst the slope parameters β would be referred to as *fixed effects*.

The following Stan code (`poisson_regression.stan`) implements a Poisson regression model with p covariates and no intercept.

```
// poisson_regression.stan

data {
    int<lower=0> n; // number of observations
    int<lower=0> p; // number of covariates
    int<lower=0> m; // number of grid points
    int<lower=0> y[n]; // response variables
    matrix[n,p] X; // matrix of covariates
    matrix[m,p] grid; // matrix of grid points
    real<lower=0> a;
    real<lower=0> b;
}
transformed data {
    real t_c = (2*a+p-1)/(2*b);
}
parameters {
    vector[p] beta;
}
model {
    sqrt(dot_self(beta)*t_c) ~ student_t(2*a, 0, 1);
    target += poisson_log_glm_lpmf( y | X, 0, beta );
}
generated quantities {
    vector[m] fn_vals;
    for (i in 1:m)
        fn_vals[i] = exp( dot_product(beta,grid[i]) );
}
```

The `generated quantities{}` block declares a vector of values for evaluating the regression function pointwise over a vector of *grid* points which are inputs in the `data{}` block.

The following PyStan code (`poisson_regression_stan.py`) simulates data from a Poisson regression model with a single covariate and then seeks to infer posterior beliefs about the value of the regression coefficient using `poisson_regression.stan`. Both here and in Sect. 8.4.2.1, the plots show the sampled data and the posterior mean regression function obtained from pointwise evaluation during posterior sampling, and then the posterior density of the single coefficient β.

```
#! /usr/bin/env python
## poisson_regression_stan.py

import stan
import numpy as np
import matplotlib.pyplot as plt

# Simulate data
gen = np.random.default_rng(seed=0)
n = 25
m = 50
T = 10
x = np.linspace(start=0, stop=T, num=n)
grid = np.linspace(start=0, stop=T, num=m)
beta = .5#gen.normal()
y = [gen.poisson(np.exp(x_i*beta)) for x_i in x]
sm_data = {'n':n, 'p':1, 'a':1, 'b':0.5, 'X':x.reshape((n,1)), 'y':y, 'm':m,
    ↪ 'grid':grid.reshape((m,1))}

# Initialise stan object
with open('poisson_regression.stan','r',newline='') as f:
    sm = stan.build(f.read(),sm_data,random_seed=1)

# Select the number of MCMC chains and iterations, then sample
chains, samples, burn = 2, 10000, 1000
fit=sm.sample(num_chains=chains, num_samples=samples, num_warmup=burn, save_warmup=False)

# Plot regression function and posterior for beta
fig,axs=plt.subplots(1,2,figsize=(10,4),constrained_layout=True)
fig.canvas.manager.set_window_title('Poisson regression posterior')
f = np.mean(fit['fn_vals'],axis=1)
true_f = [np.exp(beta*x_i) for x_i in grid]
b = fit['beta'][0]
axs[0].plot(grid,f)
axs[0].plot(grid,true_f, color='c', lw=2, linestyle='--')
axs[0].scatter(x,y, color='black')
axs[0].set_title('Posterior mean regression function')
axs[0].set_xlabel(r'$x$')
h = axs[1].hist(b,200, density=True);
axs[1].axvline(beta, color='c', lw=2, linestyle='--')
axs[1].set_title('Approximate posterior density of '+r'$\beta$')
axs[1].set_xlabel(r'$\beta$')
plt.show()
```

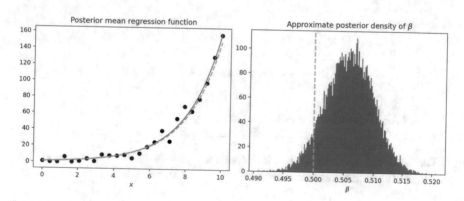

8.4.2 Logistic regression

Suppose each response $y_i \in \{0, 1\}$ is a Bernoulli indicator variable with a "success" probability (or equivalently, expected value) which is believed to be linearly dependent on $p \geq 1$ covariates $x_i \in \mathbb{R}^p$ through a logistic link function $\log\{\cdot /(1 - \cdot)\}$. These assumptions imply

$$\log \left\{ \frac{\mathbb{E}(y_i \mid x_i, \beta)}{1 - \mathbb{E}(y_i \mid x_i, \beta)} \right\} = x_i \cdot \beta,$$

for some $\beta \in \mathbb{R}^p$ and

$$y_i \mid x_i, \beta \sim \text{Bernoulli}(\{1 + \exp(-x_i \cdot \beta)\}^{-1})$$
$$\implies p(y_i \mid x_i, \beta) = \{1 + \exp((-1)^{y_i} x_i \cdot \beta)\}^{-1}.$$

8.4.2.1 Stan Implementation

The following Stan code (`logistic_regression.stan`) and PyStan code (`logistic_regression_stan.py`) implement the logistic regression model in a directly analogous way to Poisson regression in Sect. 8.4.1.1.

```
// logistic_regression.stan

data {
    int<lower=0> n; // number of observations
    int<lower=0> p; // number of covariates
    int<lower=0> m; // number of grid points
    int<lower=0,upper=1> y[n]; // response variables
    matrix[n,p] X; // matrix of covariates
    matrix[m,p] grid; // matrix of grid points
    real<lower=0> a;
    real<lower=0> b;
}
transformed data {
    real t_c = (2*a+p-1)/(2*b);
}
parameters {
    vector[p] beta;
}
model {
    sqrt(dot_self(beta)*t_c) ~ student_t(2*a, 0, 1);
    y ~ bernoulli_logit( X * beta );
}
generated quantities {
    vector[m] fn_vals;
    for (i in 1:m)

    fn_vals[i] = inv_logit( dot_product(beta,grid[i]) );
}
```

```python
#! /usr/bin/env python
## logistic_regression_stan.py

import stan
import numpy as np
import matplotlib.pyplot as plt

# Simulate data
gen = np.random.default_rng(seed=0)
n = 25
m = 50
T = 5
x = np.linspace(start=-T, stop=T, num=n)
grid = np.linspace(start=-T, stop=T, num=m)
beta = .5#gen.normal()
y = [gen.binomial(1,1/(1+np.exp(-x_i*beta))) for x_i in x]
sm_data = {'n':n, 'p':1, 'a':1, 'b':0.5, 'X':x.reshape((n,1)), 'y':y, 'm':m,
  ↪ 'grid':grid.reshape((m,1))}

# Initialise stan object
with open('logistic_regression.stan','r',newline='') as f:
    sm = stan.build(f.read(),sm_data,random_seed=1)

# Select the number of MCMC chains and iterations, then sample
chains, samples, burn = 2, 10000, 1000
fit=sm.sample(num_chains=chains, num_samples=samples, num_warmup=burn, save_warmup=False)

# Plot regression function and posterior for beta
fig,axs=plt.subplots(1,2,figsize=(10,4),constrained_layout=True)
fig.canvas.manager.set_window_title('Logistic regression posterior')
f = np.mean(fit['fn_vals'],axis=1)
true_f = [1.0/(1+np.exp(-beta*x_i)) for x_i in grid]
b = fit['beta'][0]
axs[0].plot(grid,f)
axs[0].plot(grid,true_f, color='c', lw=2, linestyle='--')
axs[0].scatter(x,y, color='black')
axs[0].set_title('Posterior mean regression function')
axs[0].set_xlabel(r'$x$')
h = axs[1].hist(b,200, density=True);
axs[1].axvline(beta, color='c', lw=2, linestyle='--')
axs[1].set_title('Approximate posterior density of '+r'$\beta$')
axs[1].set_xlabel(r'$\beta$')
plt.show()
```

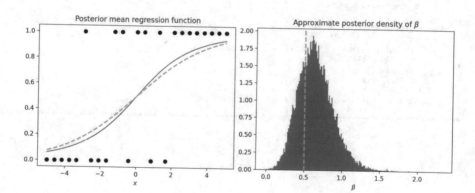

Chapter 9
Nonparametric Models

Parametric probability models provide convenient mathematical structures for approximating an individual's uncertain beliefs. For example, simple probability distributions with a small number of parameters for modelling exchangeable random quantities (Chap. 4) or a linear model for regression-exchangeable observations of a response variable (Chap. 8). The appealing simplicity of parametric models also carries a severe limitation: having assumed a parametric model, no amount of observed data can undermine the assumed certainty that the probability distribution or regression function takes that parametric form with probability one. For small sample size problems, this limitation can often seem acceptable, but for larger sample sizes the opportunity for learning potentially more complex underlying relationships grows and parametric models can become prohibitively restrictive.

More flexible modelling paradigms with the capacity to increase in complexity with increasing sample size are often referred to as *nonparametric* methods. This name can appear somewhat misleading, as these methods typically allow access to a potentially *infinite* number of parameters to provide this growth in complexity. However, the term is used to imply modelling freedom away from assuming a fixed, finite-dimensional parametric form.

The contrast between the two modelling paradigms is stark. Parametric models place probability one on a particular parametric functional form being true. Nonparametric models assume no such fixed relationship, but instead seek to spread probability mass across a much larger region of appropriate function space, such that positive mass will be assigned to arbitrarily small neighbourhoods surrounding any unknown true underlying function belonging to a much broader function class.

The higher complexity of nonparametric models can lead to a loss of analytic tractability or an increase in computational burden when performing Bayesian inference. However, there are some notable exceptions, and the next two chapters provide an overview of some popular nonparametric formulations which can be readily

The original version of this chapter has been revised due to typographic errors. The corrections to this chapter can be found at https://doi.org/10.1007/978-3-030-82808-0_12

deployed in practical applications, either for modelling probability distributions in the present chapter or regression functions in Chap. 10.

9.1 Random Probability Measures

Recall back in Chap. 2 the generalisation of De Finetti's representation theorem given in Theorem 2.2 for an infinitely exchangeable sequence $X_1, X_2, \ldots$ taking values in a space $\mathscr{X}$. Necessarily,

$$\mathbb{P}_{X_1,\ldots,X_n}(x_1, \ldots, x_n) = \int_F \prod_{i=1}^{n} F(x_i) \, dQ(F) \qquad (9.1)$$

for some probability measure Q on probability distributions F on $\mathscr{X}$.

Section 2.2 immediately proceeded to consider a parametric interpretation, with $F = F(\cdot\,; \theta)$ and $Q = Q(\theta)$ for some finite-dimensional parameter. However, a nonparametric interpretation is also possible, with Q interpreted as a probability measure over a wider class of probability distribution functions F.

The following Bayesian nonparametric models for random measures provide different specifications for the prior measure Q in (9.1), each placing mass on probability distributions with a potentially infinite number of parameters. In each case, Bayesian inference will be examined for an exchangeable sample $x_1, \ldots, x_n$ drawn from the unknown distribution F on the space $\mathscr{X}$.

9.2 Dirichlet Processes

The Dirichlet process (Ferguson 1973) is a conjugate prior for Bayesian inference about an unknown probability distribution function F.

Definition 9.1 (*Dirichlet process*) Let $\alpha > 0$ and let $\mathbb{P}_0$ be a probability measure on $\mathscr{X}$ (with distribution function F_0). A random probability measure $\mathbb{P}$ (with distribution function F) is said to be a *Dirichlet process* with *base measure* $\alpha \cdot \mathbb{P}_0$, written $\mathbb{P} \sim \mathrm{DP}(\alpha \cdot \mathbb{P}_0)$, if for every (measurable) finite partition $B_1, \ldots, B_k$ of $\mathscr{X}$,

$$(\mathbb{P}(B_1), \ldots, \mathbb{P}(B_k)) \sim \mathrm{Dirichlet}(\alpha \, \mathbb{P}_0(B_1), \ldots, \alpha \, \mathbb{P}_0(B_k)). \qquad (9.2)$$

Remark 9.1 The base measure $\mathbb{P}_0$ of the Dirichlet process is also the mean, such that for every (measurable) subset $B \subseteq \mathscr{X}$,

$$\mathbb{E}\{\mathbb{P}(B)\} = \mathbb{P}_0(B).$$

The *concentration parameter* α determines the variance

$$\mathbb{V}\{\mathbb{P}(B)\} = \frac{\mathbb{P}_0(B)\{1 - \mathbb{P}_0(B)\}}{\alpha + 1}. \qquad (9.3)$$

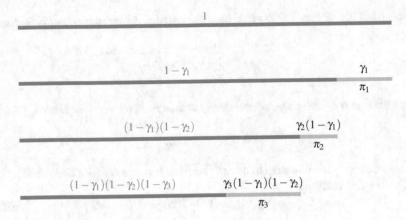

Fig. 9.1 Illustration of first three iterations of the stick-breaking process

Remark 9.2 A draw from any Dirichlet process is a discrete distribution with probability 1, even if the base measure is continuous.

Since all sampled probability distributions from $DP(\alpha \cdot \mathbb{P}_0)$ are discrete, it is possible to equivalently state the conditions for a Dirichlet process as a generative model for random probability mass functions. This generating process uses a so-called *stick-breaking* construction.

Definition 9.2 (*Stick-breaking process*) Let $\pi = (\pi_1, \pi_2, \ldots)$ be an infinite random sequence of probabilities such that $\sum_{j=1}^{\infty} \pi_j = 1$. Then π is defined as a *stick-breaking process* if

$$\pi_j = \gamma_j \prod_{k=1}^{j-1} (1 - \gamma_k), \tag{9.4}$$

where $\gamma_1, \gamma_2, \ldots$ are an infinite sequence of independent random variables in $[0, 1]$.

Definition 9.3 (*Griffiths-Engen-McCloskey distribution*) A stick-breaking process π (9.4) follows a *Griffiths-Engen-McCloskey distribution* with parameter $\alpha > 0$, written $\pi \sim GEM(\alpha)$, if $\gamma_k \sim Beta(1, \alpha)$ for all k.

Remark 9.3 The stick-breaking analogy for Definition 9.2 envisages successively breaking into pieces a stick of unit length, each time snapping off and laying down a section and then continuing to break the remaining piece of stick. For a $GEM(\alpha)$ distribution in Definition 9.3, at each break point, the proportion of remaining stick broken off and placed down follows a $Beta(1, \alpha)$ distribution. The procedure is illustrated in Fig. 9.1.

Proposition 9.1 *If $\mathbb{P} \sim \mathrm{DP}(\alpha \cdot \mathbb{P}_0)$, then the corresponding mass function satisfies*

$$\mathrm{p}(x) = \sum_{j=1}^{\infty} w_j \, \mathbb{1}_{\{\theta_j\}}(x), \tag{9.5}$$

where the atoms of mass are independently drawn from the base measure,

$$\theta_1, \theta_2, \ldots \sim \mathbb{P}_0,$$

and the masses w_j are obtained from a stick-breaking process with a Griffiths-Engen-McCloskey distribution,

$$(w_1, w_2, \ldots) \sim \mathrm{GEM}(\alpha). \tag{9.6}$$

Proof See Sethuraman (1994).

It was noted above that the Dirichlet process is a conjugate prior for an unknown probability distribution. This is now demonstrated in the following proposition.

Proposition 9.2 *Conjugacy of Dirichlet process. Suppose $\mathbf{x} = (x_1, \ldots, x_n)$ are n independent samples from $\mathbb{P}$ and $\mathbb{P} \sim \mathrm{DP}(\alpha \cdot \mathbb{P}_0)$. For ($\mathbb{P}_0$-measurable) $B \subseteq \mathscr{X}$, let*

$$\hat{\mathbb{P}}_n(B) = \frac{1}{n} \sum_{i=1}^{n} \mathbb{1}_B(x_i)$$

be the empirical measure of the samples $\mathbf{x}$, and let $\alpha_n^* = \alpha + n$, $\pi_n^* = \alpha/\alpha_n^*$ and

$$\mathbb{P}_n^*(B) = \pi_n^* \, \mathbb{P}_0(B) + (1 - \pi_n^*)\hat{\mathbb{P}}_n(B). \tag{9.7}$$

Then

$$\mathbb{P} \mid \mathbf{x} \sim \mathrm{DP}(\alpha_n^* \cdot \mathbb{P}_n^*).$$

Proof For a finite partition of (measurable) $\mathscr{X}$ subsets $B_1, \ldots, B_k$, the Dirichlet distribution prior (9.2) has density function

$$p(\mathbb{P}(B_1), \ldots, \mathbb{P}(B_k)) = \frac{\Gamma(\alpha)}{\prod_{j=1}^{k} \Gamma\{\alpha \, \mathbb{P}_0(B_j)\}} \prod_{j=1}^{k} \mathbb{P}(B_j)^{\alpha \, \mathbb{P}_0(B_j)}.$$

Let $n_j = \sum_{i=1}^{n} \mathbb{1}_{B_j}(x_i)$ be the number of samples falling inside B_j. Then the joint density of $\mathbf{x}$ is

$$p(\mathbf{x} \mid \mathbb{P}) = \prod_{j=1}^{k} \mathbb{P}(B_j)^{n_j}$$

and hence the posterior density is $\mathbf{x}$

$$p(\mathbb{P}(B_1), \ldots, \mathbb{P}(B_k) \mid \mathbf{x}) \propto p(\mathbb{P}(B_1), \ldots, \mathbb{P}(B_k)) \, p(\mathbf{x} \mid \mathbb{P})$$

$$\propto \prod_{j=1}^{k} \mathbb{P}(B_j)^{\alpha \, \mathbb{P}_0(B_j) + n_j},$$

corresponding to the Dirichlet$(\alpha_n^* \, \mathbb{P}_n^*(B_1), \ldots, \alpha_n^* \, \mathbb{P}_n^*(B_k))$ distribution.

Remark 9.4 In Proposition 9.2, as $n \to \infty$ then $\alpha_n^* \to \infty$, meaning the variance (9.3) of the Dirichlet process posterior shrinks to zero. Furthermore, the weight $\pi_n^* \to 0$ and hence the posterior mean (9.7) converges to the empirical measure, $\mathbb{P}_n^* \to \hat{\mathbb{P}}_n$.

9.2.1 Discrete Base Measure

It follows from Proposition 4.1 that independent observations from an unknown distribution function have a closed-form marginal likelihood under a Dirichlet process prior. The form of the marginal likelihood is most straightforward when the base measure $\mathbb{P}_0$ is discrete, with corresponding probability mass function p_0.

Exercise 9.1 (*Dirichlet process marginal likelihood*) Suppose $\mathbf{x} = (x_1, \ldots, x_n)$ are n independent samples from $\mathbb{P}$ and $\mathbb{P} \sim DP(\alpha \cdot \mathbb{P}_0)$. If $\mathbb{P}_0$ is discrete, show that $\mathbf{x}$ has marginal probability mass function

$$p(\mathbf{x}) = \frac{\Gamma(\alpha)}{\Gamma(\alpha + n)} \prod_{i=1}^{n} \frac{\alpha \, p_0(x_i) + \sum_{j \leq i} \mathbb{1}_{\{x_i\}}(x_j)}{\alpha \, p_0(x_i) + \sum_{j < i} \mathbb{1}_{\{x_i\}}(x_j)}. \tag{9.8}$$

Perhaps the most revealing formulation of the Dirichlet process arises from considering the predictive distribution of a further random sample.

Corollary 9.1 *The predictive distribution for a new observation x_{n+1} drawn from the same unknown distribution is*

$$p(x_{n+1} \mid \mathbf{x}) = \frac{\alpha \, p_0(x_{n+1}) + \sum_{i=1}^{n} \mathbb{1}_{\{x_{n+1}\}}(x_i)}{\alpha + n}. \tag{9.9}$$

Proof This follows immediately by expressing the predictive distribution as the ratio of the respective joint distributions (9.8) for $(\mathbf{x}, x_{n+1})$ and $\mathbf{x}$.

Remark 9.5 The form of the predictive distribution (9.9) has a clear interpretation that a further sample x_{n+1} can be viewed as a draw from the following mixture distribution: with probability $\alpha/(\alpha + n)$, a new value is sampled from the base distribution $\mathbb{P}_0$, and with the remaining probability a repeated value is sampled from the empirical distribution of values observed so far. The concentration parameter can therefore be interpreted as a notional *prior sample size* reflecting the base measure.

Remark 9.6 Following the sequential sampling procedure (9.9), the number of samples from the base distribution $\mathbb{P}_0$ follows a so-called *Chinese restaurant table distribution*. After n samples, from Teh (2010), for example, this distribution is known to have expected value

$$\alpha\{\psi_0(\alpha + n) - \psi_0(\alpha)\} \approx \alpha \log(1 + n/\alpha),$$

where $\psi_0(\cdot)$ denotes the *digamma* function, defined to be the gradient of log $\Gamma(\cdot)$.

□ **Exercise 9.2** (*Dirichlet process sampling*) Write computer code (using a language such as Python) to sample a random probability mass function from a Dirichlet process using a geometric distribution base measure with parameter 0.01. Plot three sampled probability mass functions obtained from setting $\alpha = 10, 1000, 100000$, respectively.

9.3 Polya Trees

Polya trees (Mauldin et al. 1992) are a more general class of nonparametric models for random measures which can support both continuous and discrete distributions. For real-valued random variables, Polya trees are defined on an infinite sequence of recursive partitions of a subset of the real line $B \subseteq \mathbb{R}$.

Definition 9.4 (*Binary sequences*) Let $E_0 = \emptyset$ and for $m > 1$, define

$$E_m := \{0, 1\}^m,$$
$$E := \cup_{m=0}^{\infty} E_m,$$

such that E_m is the set of all length-m *binary sequences* and E is the set of all finite-length binary sequences.

Definition 9.5 (*Binary tree of partitions*) A set $\Pi = \{\pi_0, \pi_1, \ldots\}$ of nested partitions of B is said to be a *binary tree of partitions* if $|\pi_m| = 2^m$. Clearly $\pi_0 = \{B\}$, and since the partitions are nested, the sets in each partition π_m can be indexed by elements of E_m in such a way that, for all $e \in E_{m-1}$,

$$B_e = B_{e0} \cup B_{e1}$$

for the set $B_e \in \pi_{m-1}$ and the two sets $B_{e0}, B_{e1} \in \pi_m$.

Remark 9.7 A natural binary tree of partitions of the unit interval $B = (0, 1)$ is $\pi_0 = \{(0, 1)\}$ and

$$\pi_m = \{(\sum_{j=1}^{m} e_j/2^j, 1/2^m + \sum_{j=1}^{m} e_j/2^j) \mid (e_1, \ldots, e_m) \in E_m\}, \quad m > 0,$$

illustrated in Fig. 9.2.

π_0:								B								
π_1:				B_0								B_1				
π_2:		B_{00}				B_{01}				B_{10}				B_{11}		
π_3:	B_{000}		B_{001}		B_{010}		B_{011}		B_{100}		B_{101}		B_{110}		B_{111}	
π_4:	B_{0000}	B_{0001}	B_{0010}	B_{0011}	B_{0100}	B_{0101}	B_{0110}	B_{0111}	B_{1000}	B_{1001}	B_{1010}	B_{1011}	B_{1100}	B_{1101}	B_{1110}	B_{1111}

Fig. 9.2 The first five layers of an infinite sequence of recursive partitions. The shaded regions show the path through the preceding layers to an example set B_{0110} in π_4

Remark 9.8 If $B = \mathbb{R}$ and F_0 a continuous distribution function, corresponding partitions of the real line can be obtained by applying an inverse transformation to partitions of the unit interval:

$$\pi_m = \{(F_0^{-1}(\textstyle\sum_{j=1}^{m} e_j/2^j), (F_0^{-1}(1/2^m + \sum_{j=1}^{m} e_j/2^j)) \mid (e_1, \ldots, e_m) \in E_m\}, \quad m > 0. \tag{9.10}$$

Given a binary partition $\{\pi_0, \pi_1, \ldots\}$ of B and an element $x \in B$, for each partition level m define $\epsilon_m(x)$ to be the unique length-m binary sequence $e \in E_m$ such that $x \in B_e \in \pi_m$.

⌑ **Exercise 9.3** (*Binary partition index*) Suppose an F_0-centred sequence of partitions (9.10) with $F_0(x) = \Phi(x)$, the standard normal cumulative distribution function. Evaluate $\epsilon_6(1.5)$.

Definition 9.6 (*Splitting probabilities*) Suppose $\mathbb{P}$ is a probability measure on B, and Π a binary tree of partitions of B. Let $e = (e_1, \ldots, e_m) \in E_m$ and $B_e \in \pi_m \in \Pi$. Then since $B_{e_1 \ldots e_j} \subseteq B_{e_1 \ldots e_{j-1}}$ for all $j \leq m$, the probability $\mathbb{P}(B_e)$ can be factorised as

$$\mathbb{P}(B_e) = \prod_{j=1}^{m} \mathbb{P}(B_{e_1 \ldots e_j} \mid B_{e_1 \ldots e_{j-1}}). \tag{9.11}$$

The conditional probabilities in (9.11) are known as *splitting probabilities*.

Definition 9.7 (*Polya tree*) Let Π be a binary tree of partitions and suppose $\mathscr{A} = \{\alpha_e \mid e \in E\}$ is a corresponding set of positive constants $\alpha_e > 0$ defined for all partition layers in Π. For a random probability measure $\mathbb{P}$, if for all $m > 0$ and $(e_1, \ldots, e_m) \in E_m$ the splitting probabilities satisfy

$$\mathbb{P}(B_{e_1 \ldots e_m} \mid B_{e_1 \ldots e_{m-1}}) \sim \text{Beta}(\alpha_{e_1 \ldots e_{m-1} e_m}, \alpha_{e_1 \ldots e_{m-1}(1-e_m)}),$$

then $\mathbb{P}$ is said to have a *Polya tree* distribution, written $\mathbb{P} \sim \text{PT}(\Pi, \mathscr{A})$.

Definition 9.8 (*Splitting probabilities*) The conditional probabilities in (9.11) are known as *splitting probabilities*.

Remark 9.9 The Dirichlet process from Sect. 9.2 is a special case of the Polya tree satisfying $\alpha_{e0} + \alpha_{e1} = \alpha_e$ for all $e \in E$.

Remark 9.10 Polya tree probabilities can be interpreted as products of conditional probabilities determining the path of a particle cascading down the layers of partitions, with $B \supseteq B_{e_1} \supseteq B_{e_1 e_2} \supseteq \ldots$. For example, in Fig. 9.2, the probability of the highlighted set B_{0110} is obtained through a product,

$$\mathbb{P}(B_{0110}) = \mathbb{P}(B_0)\,\mathbb{P}(B_{01} \mid B_0)\,\mathbb{P}(B_{011} \mid B_{01})\,\mathbb{P}(B_{0110} \mid B_{011}),$$

where each term (splitting probability) has an independent Beta distribution with parameters corresponding to that path.

The specification of a binary tree of partitions according to a base probability measure (9.10) allows the Polya tree distribution to be easily centred around that distribution.

Proposition 9.3 *For a chosen a probability measure* $\mathbb{P}_0$ *with distribution function* F_0, *suppose* $\mathbb{P} \sim \mathrm{PT}(\Pi, \mathscr{A})$ *where each* $\pi_m \in \Pi$ *satisfies* (9.10). *If the positive constants* $\mathscr{A}$ *are chosen to be symmetric such that for all* $e \in E$, $\alpha_{e0} = \alpha_{e1}$ *then* $\mathbb{E}_{\mathbb{P}} = \mathbb{P}_0$.

Proof If, for all $e, \alpha_{e0} = \alpha_{e1}$ then by symmetry, for all $m > 0$ and $e \in E_m$, $\mathbb{E}\{\mathbb{P}(B_e)\} = 1/2^m$. The result follows from the usual inversion rule for continuous distribution functions.

The conjugacy of the Polya tree prior follows immediately from the conjugacy of the beta distribution for Bernoulli observations noted in Table 4.1 of Sect. 4.2.

Proposition 9.4 *Conjugacy of Polya tree. Suppose* $\mathbf{x} = (x_1, \ldots, x_n)$ *are* n *independent samples from* $\mathbb{P}$ *and* $\mathbb{P} \sim \mathrm{PT}(\Pi, \mathscr{A})$. *For* $e \in E$, *let*

$$n_e = \sum_{i=1}^{n} \mathbb{1}_{B_e}(x_i)$$

be the number of samples which fall inside the set B_e, *and let* $\mathscr{A}_n^* = \{\alpha_e + n_e \mid e \in E\}$. *Then*

$$\mathbb{P} \mid \mathbf{x} \sim \mathrm{PT}(\Pi, \mathscr{A}_n^*).$$

Proof For each sample x_i and each non-trivial partition level $m > 0$, recall $\epsilon_{m-1}(x_i)$ as the unique binary sequence of length $m - 1$ such that $x_i \in B_{\epsilon_{m-1}(x_i)}$. Conditional on $\epsilon_{m-1}(x_i)$, x_i must fall in either $B_{\epsilon_{m-1}(x_i)0}$ or $B_{\epsilon_{m-1}(x_i)1}$; from these two possibilities, x_i falls in $B_{\epsilon_{m-1}(x_i)e_m}$ with an unknown, $\mathrm{Beta}(\alpha_{\epsilon_{m-1}(x_i)e_m}, \alpha_{\epsilon_{m-1}(x_i)(1-e_m)})$ distributed probability. Denoting this probability θ,

$$p(\theta \mid x_i \in B_{\epsilon_{m-1}(x_i)e_m}) \propto p(x_i \in B_{\epsilon_{m-1}(x_i)e_m} \mid \theta) \times p(\theta)$$
$$\propto \theta \times \theta^{\alpha_{\epsilon_{m-1}(x_i)e_m}-1} (1-\theta)^{\alpha_{\epsilon_{m-1}(x_i)(1-e_m)}-1}$$
$$= \theta^{\epsilon_{m-1}(x_i)e_m} (1-\theta)^{\alpha_{\epsilon_{m-1}(x_i)(1-e_m)}-1},$$

which is proportional to the density of $\text{Beta}(\alpha_{\epsilon_{m-1}(x_i)e_m} + 1, \alpha_{\epsilon_{m-1}(x_i)(1-e_m)})$. The result follows.

9.3.1 Continuous Random Measures

As noted above, Polya trees can be constructed to give probability one to either discrete or continuous distributions. The special case of the Dirichlet process obtained when $\alpha_{e0} + \alpha_{e1} = \alpha_e$ for all e exemplifies the discrete case. For guaranteeing continuous probability distributions, Lavine (1992) showed that "as long as the α_e's do not decrease too rapidly with m", $\mathbb{P}$ will be continuous; a commonly used choice is $\alpha_{e_1 \dots e_m} = \alpha m^2$ for some single parameter $\alpha > 0$.

As with the Dirichlet process, it follows from Proposition 4.1 that independent observations from an unknown distribution function have a closed-form marginal likelihood under a Polya tree prior; in this case, this marginal likelihood is most straightforward when the base measure $\mathbb{P}_0$ is continuous.

Proposition 9.5 *Polya tree marginal likelihood. Suppose* $\mathbf{x} = (x_1, \dots, x_n)$ *are* n *independent samples from* $\mathbb{P}$ *and* $\mathbb{P} \sim \text{PT}(\Pi, \mathscr{A})$. *If* $\mathbb{P}_0 = \mathbb{E}(\mathbb{P})$ *is continuous with corresponding probability density function* $\mathbb{p}_0$, *then* $\mathbf{x}$ *has marginal probability density function*

$$p(\mathbf{x}) = \prod_{i=1}^{n} \mathbb{p}_0(x_i) \prod_{j=2}^{n} \prod_{m=1}^{m^*(\mathbf{x},j)} \frac{(\alpha_{\epsilon_m(x_j)} + n_{\epsilon_m(x),j})(\alpha_{\epsilon_{m-1}(x_j)0} + \alpha_{\epsilon_{m-1}(x_j)1})}{\alpha_{\epsilon_m(x_j)}(\alpha_{\epsilon_{m-1}(x_j)0} + \alpha_{\epsilon_{m-1}(x_j)1} + n_{\epsilon_{m-1}(x),j})},$$

where $n_{e,j} = \sum_{i<j} \mathbb{1}_{B_e}(x_i)$ *and* $m^*(\mathbf{x}, j) = \min\{m > 0 \mid \epsilon_m(x_i) \neq \epsilon_m(x_j), i < j\}$ *is the highest partition level for which none of the first* $(j-1)$ *samples in* $\mathbf{x}$ *lie within the same set as* x_j.

Proof See Berger and Guglielmi (2001).

□ **Exercise 9.4** (*Polya tree sampling*) Write computer code (using a language such as Python) to sample a random probability density function from a Polya tree model with $\alpha_{e_1 \dots e_m} = \alpha m^2$ and a binary tree of partitions Π centred on $F_0(x) = \Phi(x)$. Plot three sampled densities obtained from setting $\alpha = 1, 100, 10000$, respectively.

9.4   Partition Models

A Polya tree defines a random probability measure $\mathbb{P}$ on a fixed collection of nested partitions of a space B, specifying consistent probabilities at each layer. *Partition models* are somewhat simpler, specifying $\mathbb{P}$ on a single, unknown partition π.

For each B-subset of the partition π, a relatively simple statistical model is typically assumed. The nonparametric flexibility of a partition model comes from allowing uncertainty about the partition to extend to the dimension $|\pi|$; by not assuming an upper bound for the size of the partition, $\mathbb{P}$ can assume a potentially infinite number of parameters. A simple analogy is approximating an arbitrarily complex function with a step function with an unlimited number of steps.

9.4.1   *Partition Models: Bayesian Histograms*

For simplicity of exposition assume $B = [a, b] \subset \mathbb{R}$ is an interval on the real line, and that $\mathbb{P}$ is an unknown continuous probability measure on $[a, b]$ with density $\mathbb{p}$. A histogram on $[a, b]$ can be viewed as a partition model: the interval $[a, b]$ is divided into bins by a sequence of $m \geq 0$ cut points τ, where $\tau = \emptyset$ when $m = 0$ and otherwise $\tau = (\tau_1, \ldots, \tau_m)$ with $a \equiv \tau_0 < \tau_1 < \ldots < \tau_m < \tau_{m+1} \equiv b$. The cut points define a corresponding partition $\pi_\tau = \{B_1, \ldots, B_{m+1}\}$ where $B_j = [\tau_{j-1}, \tau_j)$ is the jth bin of the histogram.

A histogram assumes constant density within each bin, leading to an overall piecewise constant density on $[a, b]$ with m steps. Leonard (1973) and Gelman et al. (2013, p. 545) presented the following Bayesian model for such a density.

Definition 9.9 (*Bayesian histogram*) Let $\alpha > 0$ and let $\mathbb{P}_0$ be a probability measure on $[a, b]$. Let τ be an increasing sequence of m cut points partitioning $[a, b]$ into $(m + 1)$ segments, with corresponding segment probabilities $\theta = (\theta_1, \ldots, \theta_{m+1})$ satisfying $\sum_{j=1}^{m+1} \theta_j = 1$. A *Bayesian histogram* model for a random probability measure $\mathbb{P}$ assumes the following representation for the density $\mathbb{p}$:

$$\mathbb{p}(x \mid m, \tau, \theta) = \sum_{j=1}^{m+1} \mathbb{1}_{[\tau_{j-1}, \tau_j)}(x) \, \frac{\theta_j}{\tau_j - \tau_{j-1}},$$

$$\theta \mid m, \tau \sim \text{Dirichlet}\{\alpha \, \mathbb{P}_0([\tau_0, \tau_1)), \ldots, \alpha \, \mathbb{P}_0([\tau_m, \tau_{m+1}))\}. \tag{9.12}$$

Remark 9.11 The base probability measure $\mathbb{P}_0$ from Definition 9.9 is the prior expectation for $\mathbb{P}$, such that for ($\mathbb{P}_0$-measurable) $A \subseteq [a, b]$, $\mathbb{E}\{\mathbb{P}(A)\} = \mathbb{P}_0(A)$.

Given samples $x_1, \ldots, x_n$ from an unknown continuous probability distribution $\mathbb{P}$, the Bayesian histogram model (9.12) provides another conjugate model for $\mathbb{P}$ with closed-form marginal likelihood.

Proposition 9.6 (Bayesian histogram marginal likelihood) *Suppose* $\mathbf{x} = (x_1, \ldots, x_n)$ *are n independent samples from an unknown continuous distribution* $\mathbb{P}$ *with density defined by (9.12). Conditional on m and* τ, *the posterior distribution of* θ *given* $\mathbf{x}$ *is*

$$\theta \mid m, \tau, \mathbf{x} \sim \text{Dirichlet}\{\alpha \, \mathbb{P}_0([\tau_0, \tau_1) + n_1), \ldots, \alpha \, \mathbb{P}_0([\tau_m, \tau_{m+1}) + n_{m+1})\} \quad (9.13)$$

and $\mathbf{x}$ *has marginal probability density function*

$$p(\mathbf{x} \mid m, \tau) = \frac{\Gamma(\alpha)}{\Gamma(\alpha + n)} \prod_{j=1}^{m+1} \frac{\Gamma\{\alpha \, \mathbb{P}_0([\tau_{j-1}, \tau_j)) + n_j\}}{\Gamma\{\alpha \, \mathbb{P}_0([\tau_{j-1}, \tau_j))\}(\tau_j - \tau_{j-1})^{n_j}}, \quad (9.14)$$

where $n_j = \sum_{i=1}^{n} \mathbb{1}_{[\tau_{j-1}, \tau_j)}(x_i)$ *is the number of samples lying in the segment* $[\tau_{j-1}, \tau_j)$.

Proof The likelihood function for (9.12) is

$$p(\mathbf{x} \mid m, \tau, \theta) = \prod_{j=1}^{m+1} \left(\frac{\theta_j}{\tau_j - \tau_{j-1}} \right)^{n_j}$$

and the results simply follow from the conjugacy of the multinomial and Dirichlet distributions (*cf.* Table 4.1).

Remark 9.12 Assuming α to be relatively small, the marginal likelihood (9.14) is highest when the bin counts $\{n_j\}$ are each either very large or very small. Therefore equal bin counts would not correspond to a good partition; these would be better modelled with a single bin.

To complete the nonparametric formulation of the Bayesian histogram, a prior distribution must be assigned to the number and location of the cut points, (m, τ). The canonical choice for this assignment is a Poisson process on $[a, b]$ with rate $\nu > 0$ for the arrivals of cut points, leading to a joint prior density

$$p(m, \tau) = \nu^m \exp\{-\nu(b - a)\}. \quad (9.15)$$

For any choice of prior density $p(m, \tau)$, the corresponding posterior density for the number of cut points is given up to proportionality by

$$p(m, \tau \mid \mathbf{x}) \propto p(m, \tau) \frac{\Gamma(\alpha)}{\Gamma(\alpha + n)} \prod_{j=1}^{m+1} \frac{\Gamma\{\alpha \, \mathbb{P}_0([\tau_{j-1}, \tau_j)) + n_j\}}{\Gamma\{\alpha \, \mathbb{P}_0([\tau_{j-1}, \tau_j))\}(\tau_j - \tau_{j-1})^{n_j}}. \quad (9.16)$$

Estimation of posterior expectations taken with respect to (9.16) can straightforwardly proceed using (reversible jump) Markov chain Monte Carlo sampling (Green 1995) (*cf.* Chap. 5).

9.4.2 Bayesian Histograms with Equal Bin Widths

Now consider three simplifications of the histogram model (9.12). First, for simplicity of presentation and without loss of generality, suppose that the interval of interest is the unit interval $B = [0, 1]$.

Second, suppose the base measure $\mathbb{P}_0$ in Definition 9.9 is the natural default choice for the unit interval, *Lebesgue measure*, such that $\mathbb{P}_0([\tau_{j-1}, \tau_j)) = \tau_j - \tau_{j-1}$.

Third, suppose the unknown distribution $\mathbb{P}$ is characterised by a partition model with an unknown number of equally spaced cut points on $[0, 1]$. To simplify subsequent notation, let m now denote the number of equally sized segments rather than the number of cut points. Making this assumption is then equivalent to specifying $p(m, \tau)$ through a non-degenerate probability model $p(m)$ for the unknown number of segments, whilst for $m > 1$ the conditional distribution $p(\tau \mid m)$ assigns probability one to the vector τ^* with jth element

$$\tau_j^* = \frac{j}{m}, \quad j = 1, \ldots, m - 1.$$

With these three conditions, the posterior density (9.16) simplifies to

$$p(m \mid \mathbf{x}) \propto p(\mathbf{x}, m) = \frac{p(m) \, m^n \, \Gamma(\alpha)}{\Gamma(\alpha + n) \, \Gamma(\alpha/m)^m} \prod_{j=1}^{m} \Gamma\left\{\frac{\alpha}{m} + n_j^{(m)}\right\}, \qquad (9.17)$$

where $n_j^{(m)}$ is the number of samples lying between $\frac{j-1}{m}$ and $\frac{j}{m}$.

Also, under this simplified model it follows from (9.13) that after marginalising θ, the posterior predictive density conditional on m satisfies

$$\mathbb{p}(x \mid m, \mathbf{x}) = \sum_{j=1}^{m} \mathbb{1}_{[j-1,j)}(m \, x) \, \frac{\alpha + m \, n_j^{(m)}}{\alpha + n}. \qquad (9.18)$$

Model averaging (9.18) with respect to the prior distribution for m obtains the marginal predictive density

$$\mathbb{p}(x \mid \mathbf{x}) = \sum_{m=1}^{\infty} \frac{p(m) \, m^n \, \Gamma(\alpha) \prod_{j'=1}^{m} \Gamma\left\{\alpha/m + n_{j'}^{(m)}\right\}}{\Gamma(\alpha + n) \, \Gamma(\alpha/m)^m} \sum_{j=1}^{m} \mathbb{1}_{[j-1,j)}(m \, x) \, \frac{\alpha + m \, n_j^{(m)}}{\alpha + n}.$$
$$(9.19)$$

The predictive density (9.19) could be estimated by a finite approximation of the outer sum.

Remark 9.13 Relaxing the first two assumptions of this section and returning to a general base measure $\mathbb{P}_0$ on $[a, b]$ with corresponding distribution function F_0, the same principle of equal bin-width histogram modelling could equally be applied on the F_0-scale, such that segment j is the interval $[\tau_{j-1}^*, \tau_j^*]$ where $\tau_j^* = F_0^{-1}(j/m)$. This is somewhat analogous to the Polya tree partition of (9.10). Then, for example,

$$p(m \mid \mathbf{x}) \propto \frac{p(m)\ \Gamma(\alpha)}{\Gamma(\alpha+n)\ \Gamma(\alpha/m)^m} \prod_{j=1}^{m}(\tau_j^* - \tau_{j-1}^*)^{n_j^{(m)}} \Gamma\left\{\frac{\alpha}{m} + n_j^{(m)}\right\}.$$

9.4.2.1 Approximate Inference

The simplicity of inference for the Bayesian histogram with equal bin widths (and using Lebesgue measure as the base measure) was illustrated by the joint density $p(\mathbf{x}, m)$ (9.17). With just a single unknown parameter m, it is feasible to take a finite sum approximation of the marginal likelihood,

$$p(\mathbf{x}) = \sum_{m=1}^{\infty} p(\mathbf{x}, m), \qquad (9.20)$$

by terminating the summation at a suitably large value of m. A useful approximation of $p(m \mid \mathbf{x})$ is thereby obtained from the ratio of (9.17) and (9.20). Access to the discrete posterior distribution for m allows direct posterior inference without resorting to computational methods (*cf.* Chap. 5). For example, it is straightforward to calculate posterior expectations for functions of interest as simple weighted sums.

Assuming a relatively uninformative geometric prior distribution for m, the following Python code (`bayesian_histogram.py`) obtains the maximum a posteriori value of m in this setting, and more importantly illustrates model averaging (*cf.* Sect. 7.2), marginalising over m to obtain the posterior expectation of the unknown density function.

```
## bayesian_histogram.py

import numpy as np
from scipy.special import gammaln
from collections import Counter

def log_prior(m,theta=.01):
    return(np.log(theta)+(m-1)*np.log(1-theta))

def log_likelihood(x,m,a=0.0,b=1.0,alpha=10):
    c = Counter([int(x_i*m/(b-a)) for x_i in x-a])
    ml = len(x)*np.log(m/(b-a)) - len(c)*gammaln(alpha/m)
    ml += sum([gammaln(alpha/m + c[i]) for i in c])
    return(ml,c)

def density(t,a,b,m,c,n):
    width = (b-a)/m
    ind = min(int((t-a)/(b-a)*m),m-1)
    return((c[ind] + width)/(b - a + n)/width)

def model_average(x,max_m=100,grid=100,a=0.0,b=1.0,alpha=10):
    n = len(x)
    max_post = -float('inf')
    sum_probs = 0
    ave_density = np.zeros(grid)
    probs = np.zeros(max_m+1)
    for m in range(1,max_m+1):
        log_lhd,ctr = log_likelihood(x,m)
        log_post= log_lhd + log_prior(m)
        if log_post > max_post:
            renormalise = np.exp(max_post-log_post)
            sum_probs *= renormalise
            probs[:m] *= renormalise
            ave_density *= renormalise
            max_post = log_post
            max_m = m
            max_m_ctr = ctr
        probs[m] = np.exp(log_post-max_post)
        sum_probs += probs[m]
        for i in range(grid):
            ave_density[i] += probs[m] * density(a+(b-a)*i/(grid-1.0),a,b,m,ctr,n)

    ave_density /= sum_probs;
    return(ave_density, max_m, max_m_ctr, probs/sum_probs)
```

Inference under this model is illustrated by the following Python code (`bayes_histogram_simulate.py`), where 10,000 observations are simulated from a mixture of two beta distributions. The three plots generated by the code show the true mixture density (dashed line) and the model-averaged posterior expected density (solid line), the approximated posterior distribution for m, and the histogram density which contributed most to the model-averaged density function, corresponding to the maximum a posteriori value of m which was equal to 27.

```python
#! /usr/bin/env python
## bayes_histogram_simulate.py
from bayesian_histogram import *
import matplotlib.pyplot as plt
from scipy.stats import beta
gen = np.random.default_rng(seed=0)

def simulate_beta_mixture(n, beta_pars, probs):
    z=gen.choice(len(probs), n, p=probs)
    return(np.array([gen.beta(*(beta_pars[z_i])) for z_i in z]))
def mixture_density(x, beta_pars, probs):
    return(sum([probs[i]*beta.pdf(x,*beta_pars[i]) for i in range(len(probs))]))

n = 10000
beta_pars=[[20,10],[2,3]]
probs=[0.3,0.7]
x = simulate_beta_mixture(n, beta_pars, probs)
ave_density, m, ctr, pm = model_average(x,40)
m_density = [(ctr[i]*m + 1)/(1 + n) for i in range(m)]

fig,axs=plt.subplots(1,3,figsize=(12,4),constrained_layout=True)
fig.canvas.manager.set_window_title('Bayes histogram posterior')
grid = np.linspace(start=0, stop=1, num=len(ave_density))
true_f = [mixture_density(t, beta_pars, probs) for t in grid]
axs[0].plot(grid,ave_density)
axs[0].plot(grid,true_f, color='c', lw=2, linestyle='--')
axs[1].bar(range(len(pm)),pm)
axs[1].set_xlabel(r'$m$')
axs[1].set_title('Posterior '+r'$p(m\vert x,y)$')
axs[0].set_title('Averaged posterior mean density function')
axs[2].autoscale(enable=True, axis='x', tight=True)
axs[2].step(np.linspace(start=0, stop=1, num=m),m_density,where='post')
axs[2].set_title('Posterior mean density for '+r'$m=$'+str(m))
for plt_ind in [0,2]:
    axs[plt_ind].autoscale(enable=True, axis='x', tight=True)
    axs[plt_ind].set_xlabel(r'$x$')
    axs[plt_ind].set_ylim(bottom=0)
plt.show()
```

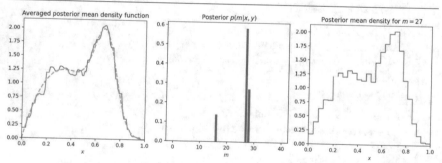

Chapter 10
Nonparametric Regression

Chapter 9 introduced the concept of nonparametric modelling, with a focus on infinite-dimensional parameter representations for unknown probability measures. In this chapter, attention turns to regression modelling, introduced in Chap. 8 with the linear model.

10.1 Nonparametric Regression Modelling

Recall from Chap. 8 the regression problem of expressing probabilistic beliefs about real-valued response variables $y_1, y_2, \ldots$ which are thought to statistically depend on corresponding known p-dimensional covariates $x_1, x_2, \ldots$. Regression exchangeability (Definition 8.1) of the response-covariate pairs (y_i, x_i) was noted to be the natural extension of standard exchangeability in this setting, which simplifies the regression task to learning a common parametric form for the conditional distribution of $y_i \mid x_i$.

The linear model was shown in Chap. 8 to be a highly flexible parametric likelihood model in the regression-exchangeable framework. The models presented in this chapter provide further flexibility through infinite-dimensional representations of regression functions $f : x \mapsto \mathbb{E}(y \mid x)$; these will include natural extensions of the linear model which have unbounded dimension.

As with nonparametric probability models, nonparametric regression models should allow arbitrarily close representations of functions f from a wider class of regression functions than fixed parametric forms allow. The first such example considered is the Gaussian process, popularised for its analytical tractability and its close relationship with the linear model.

© The Author(s), under exclusive license to Springer Nature Switzerland AG 2021
N. Heard, *An Introduction to Bayesian Inference, Methods and Computation*,
https://doi.org/10.1007/978-3-030-82808-0_10

10.2 Gaussian Processes

Consider the standard regression problem of making inference about an unknown function $f : \mathscr{X} \to \mathbb{R}$ defined on a space $\mathscr{X} \subseteq \mathbb{R}^p$ for some $p \geq 1$. A Gaussian process prior distribution for f assumes a multivariate normal distribution for the function values $f(\mathbf{x}) = (f(x_1), \ldots, f(x_n))$ at any finite collection of points $\mathbf{x} = (x_1, \ldots, x_n)$ in $\mathscr{X}$, according to the following specification.

Definition 10.1 (*Kernel function*) A symmetric function $k : \mathscr{X} \times \mathscr{X} \to \mathbb{R}$ is a *positive semidefinite kernel* if, for all $x_1, \ldots, x_n \in \mathscr{X}$, and $c_1, \ldots, c_n \in \mathbb{R}$,

$$\sum_{i=1}^{n} \sum_{j=1}^{n} c_i \, c_j \, k(x_i, x_j) \geq 0.$$

If $k(x, x')$ is a function of $x - x'$ then the kernel is said to be *stationary*; if $k(x, x')$ is a function of $|x - x'|$ then the kernel is *isotropic*.

Example 10.1 (*Example kernel functions*) The following examples satisfy the positive semidefinite requirement of Definition 10.1 for $\alpha, \rho > 0$.

- Dot product/linear:

$$k(x, x') = \alpha^2 \, (x \cdot x').$$

- Squared exponential/radial basis function:

$$k(x, x') = \alpha^2 \, \exp(-.5 \, \|x - x'\|^2 / \rho^2). \tag{10.1}$$

- Exponential:

$$k(x, x') = \alpha^2 \, \exp(-\|x - x'\| / \rho). \tag{10.2}$$

Definition 10.2 (*Gaussian process*) Let $m : \mathscr{X} \to \mathbb{R}$ be any function and $k : \mathscr{X} \times \mathscr{X} \to \mathbb{R}$ be a positive semidefinite kernel. Then $\{f(x) \mid x \in \mathscr{X}\}$ is a *Gaussian process* with mean function m and *covariance function* k, written $f \sim \mathrm{GP}(m, k)$ if for any $\mathbf{x} = (x_1, \ldots, x_n)$,

$$f(\mathbf{x}) \sim \mathrm{Normal}_n \left(m(\mathbf{x}), K(\mathbf{x}, \mathbf{x}) \right),$$

where

$$K(\mathbf{x}, \mathbf{x}) = \begin{bmatrix} k(x_1, x_1) \ k(x_1, x_2) \ \ldots \ k(x_1, x_n) \\ k(x_2, x_1) \ k(x_2, x_2) \ \ldots \ k(x_2, x_n) \\ \vdots \qquad \vdots \qquad \ddots \qquad \vdots \\ k(x_n, x_1) \ k(x_n, x_2) \ \ldots \ k(x_n, x_n) \end{bmatrix}. \tag{10.3}$$

Remark 10.1 The squared exponential kernel (10.1) is the most commonly used kernel in Gaussian process modelling; samples from processes with this kernel are infinitely differentiable (Rasmussen and Williams 2005, p. 83). For the exponential kernel (10.2), samples are continuous but not differentiable (Rasmussen and Williams 2005, p. 86).

Exercises 10.1 (*Gaussian process closure*) Suppose $f \sim \mathrm{GP}(m, k)$ and $m \sim \mathrm{GP}(m_0, k_0)$, where m_0 is any function and k and k_0 are positive semidefinite kernels. Show that marginally,

$$f \sim \mathrm{GP}(m_0, k + k_0). \tag{10.4}$$

10.2.1 Normal Errors

Available information about the function is commonly assumed to be limited to a finite number of pointwise, typically noisy, real-valued observations $\mathbf{y} = (y_1, \ldots, y_n)$ of the function values at domain points $\mathbf{x} = (x_1, \ldots, x_n)$ in $\mathcal{X}$. If those observations can be assumed to satisfy

$$y_i = f(x_i) + \varepsilon_i, \tag{10.5}$$

where the observation errors $(\varepsilon_1, \ldots, \varepsilon_n)$ are independent $\mathrm{Normal}(0, \sigma^2)$ variables, then the Gaussian process is a conjugate prior for f.

Proposition 10.1 (Conjugacy of Gaussian process) *If independently $y_i \sim \mathrm{Normal}(f(x_i), \sigma^2)$, $i = 1, \ldots, n$ and $f \sim \mathrm{GP}(m, k)$, then the posterior distribution for f is again a Gaussian process*

$$f \mid \mathbf{x}, \mathbf{y} \sim \mathrm{GP}(m^*, k^*),$$

where

$$m^*(x) = m(x) + k(x, \mathbf{x})\{K(\mathbf{x}, \mathbf{x}) + \sigma^2 I_n\}^{-1}(\mathbf{y} - m(\mathbf{x})),$$
$$k^*(x, x') = k(x, x') - k(x, \mathbf{x})\{K(\mathbf{x}, \mathbf{x}) + \sigma^2 I_n\}^{-1}k(\mathbf{x}, x'). \tag{10.6}$$

Proof The result follows from the conjugacy of normal-normal mixtures exploited in Chap. 8.

It follows from Proposition 4.1 that observations of an unknown function with independent Gaussian errors have a closed-form marginal likelihood under a Gaussian process prior.

Proposition 10.2 (Gaussian process marginal likelihood) *If $f \sim \mathrm{GP}(m, k)$ and independently $y_i \sim \mathrm{Normal}(f(x_i), \sigma^2)$, $i = 1, \ldots, n$, then $\mathbf{y} \mid \mathbf{x}$ has a marginal likelihood satisfying*

$$p(\mathbf{y} \mid \mathbf{x}, k, m, \sigma) = \frac{\exp\left[-(\mathbf{y} - m(\mathbf{x}))^{\mathsf{T}}\{K(\mathbf{x}, \mathbf{x}) + \sigma^2 I_n\}^{-1}(\mathbf{y} - m(\mathbf{x}))/2\right]}{|K(\mathbf{x}, \mathbf{x}) + \sigma^2 I_n|^{\frac{1}{2}}(2\pi)^{\frac{n}{2}}}. \tag{10.7}$$

Proof As noted in Rasmussen and Williams (2005, p. 19), the likelihood (10.7) is obtained directly from observing that $\mathbf{y} \mid \mathbf{x} \sim \text{Normal}_n(m(\mathbf{x}), K(\mathbf{x}, \mathbf{x}) + \sigma^2 I_n)$.

In a further duality with the linear model, the inverse-gamma distribution can provide a conjugate prior distribution for the error variance σ^2 if the kernel k can be satisfactorily factorised as $k(x, x') = \sigma^2 k'(x, x')$ for a kernel k', such that beliefs about the parameters of k' do not depend on σ.

Corollary 10.1 *Under the conditions of Proposition 10.2, suppose $k(x, x') = \sigma^2 k'(x, x')$ and correspondingly the matrix $K(\mathbf{x}, \mathbf{x}) = \sigma^2 K'(\mathbf{x}, \mathbf{x})$. Assuming the conjugate prior,*

$$\sigma^{-2} \sim \text{Gamma}(a, b),$$

a further marginalisation of the likelihood (10.7) is

$$p(\mathbf{y} \mid \mathbf{x}, k, m) = \frac{1}{(2\pi)^{\frac{n}{2}} |K'(\mathbf{x}, \mathbf{x}) + I_n|^{\frac{1}{2}}} \cdot \frac{\Gamma(a_n) \, b^a}{\Gamma(a) \, b_n^{a_n}}, \tag{10.8}$$

where

$$a_n = a + n/2,$$
$$b_n = b + (\mathbf{y} - m(\mathbf{x}))^\top \{K'(\mathbf{x}, \mathbf{x}) + I_n\}^{-1}(\mathbf{y} - m(\mathbf{x}))/2.$$

Hence (cf. Proposition 8.2),

$$\mathbf{y} \mid \mathbf{x}, k, m \sim \text{St}_n(2a, m(\mathbf{x}), b(K'(\mathbf{x}, \mathbf{x}) + I_n)/a).$$

Exercises 10.2 (*Linear model as a Gaussian process*) Conditional on σ, express the Bayes linear model with simplified conjugate prior (Sect. 8.2.1),

$$y \sim \text{Normal}_n(X\beta, \sigma^2 I_n),$$
$$\beta \sim \text{Normal}_p(0, \sigma^2 \lambda^{-1} I_p),$$

as normal error observations (10.5) of a Gaussian process.

10.2.2 Inference

With normally distributed observation errors leading to closed-form expressions for the marginal likelihood in Proposition 10.2 and Corollary 10.1, inferential attention is often primarily focused on the selection of the covariance kernel and the associated parameters, and secondarily on the mean function (which is often simply assumed to be zero everywhere).

Remark 10.2 There are two related reasons why a zero mean might safely be assumed, without significant loss of generality, for Gaussian process modelling of an unknown function f.

First, if a non-zero mean function $m(x)$ is assumed to be known, then attention can switch to quantifying uncertainty about the deviation $(f - m) \sim GP(0, k)$; inference about $f - m$ is then based on correspondingly detrended observations (x_i, y_i') where $y_i' = y_i - m(x_i)$, $i = 1, \ldots, n$.

Second, if the mean function $m(x)$ is unknown but can be assumed to also have a Gaussian process prior with known mean function $m_0(x)$, then by Exercise 10.1 the marginal distribution for f is again a Gaussian process with mean $m_0(x)$ and an additively modified covariance kernel (10.4). The known mean function $m_0(x)$ could be subtracted from the observation process and inference about $f - m_0$ could again proceed with the assumption of a zero-mean Gaussian process.

For inference on covariance kernel parameters, there are no further analytical results and computational inferential methods are required, such as Markov chain Monte Carlo (Sect. 5.3). Fortunately, implementation in the probabilistic programming language Stan (Sect. 6.2) is straightforward, as demonstrated by the following synthetic regression data example.

Example 10.2 Consider the normal error model from Corollary 10.1 which assumes an inverse-gamma prior for σ^2, and suppose the assumed covariance function is the popular squared exponential covariance kernel (10.1). For simplicity of presentation, the following Stan code (`gp_regression.stan`) assumes a univariate unknown function following a Gaussian process with zero mean function and uninformative, improper priors for the amplitude parameter α and the length-scale parameter ρ which determine the squared exponential kernel.

```
// gp_regression.stan

functions {
    real mean_fn(real[] t,real[] x, vector ky, int n,real a, real r){
       return dot_product(cov_exp_quad(t, x, a, r)[1],ky);
    }
}
data {
    int<lower=0> n; // number of observations
    int<lower=0> m; // number of grid points
    vector[n] y; // response variables
    real x[n]; // vector of covariates
    real grid[m]; // vector of grid points
    real<lower=0> a;
    real<lower=0> b;
}
transformed data {
    vector[n] mu = rep_vector(0, n);
}
parameters {
    real<lower=0> alpha; //kernel amplitude
    real<lower=0> rho; //kernel lengthscale
}
transformed parameters {
    matrix[n,n] Sigma = cov_exp_quad(x, alpha, rho);
    for (i in 1:n)
         Sigma[i,i] += 1;
}
model {
    y ~ multi_student_t(2*a, mu, b/a * Sigma);
}
generated quantities {
    vector[n] Ky = inverse_spd(Sigma) * y;
    vector[m] fn_vals;
    for (i in 1:m)
        fn_vals[i] = mean_fn(segment(grid,i,1),x,Ky,n,alpha,rho);
}
```

Notably, Stan has an in-built squared exponential covariance function, `cov_exp_quad`, which is used twice in the code: first within the `functions{}`

block for placing user-defined functions, where the covariance function is required for obtaining the posterior mean regression function (10.6), and second within the `transformed parameters{}` block for calculating the covariance matrix factor $(K'(\mathbf{x}, \mathbf{x}) + I_n)$ needed for evaluating the likelihood (10.8). For the latter use case, the observation in Corollary 10.1 that y is marginally multivariate Student's t-distributed is utilised to obtain a very simple statement for the `model{}` block.

The following PyStan code (`gp_regression_stan.py`) simulates univariate functional data with independent standard normal errors, where the true underlying function is

$$f(x) = 10 + 5 \sin(x) + \frac{x^2}{5}, \quad 0 \le x \le 10. \tag{10.9}$$

The code then calls `gp_regression.stan` in order to make posterior inference about the two parameters of the squared exponential kernel. Two inferential summary plots are provided for illustration: First, the posterior mean regression function, evaluated at 50 equally spaced grid points; second, the posterior distribution of the most interesting length-scale parameter ρ, which determines the smoothness of the regression function by controlling the rate at which covariance decreases with increasing distance between input points.

```python
#! /usr/bin/env python
## gp_regression_stan.py

import stan
import numpy as np
import matplotlib.pyplot as plt

def reg_fn(t): return (10+5*np.sin(t)+t**2/5.0)

# Simulate data
gen = np.random.default_rng(seed=0)
n = 40
m = 50
T = 10
x = np.linspace(start=0, stop=T, num=n)
y = [gen.normal(loc=reg_fn(x_i)) for x_i in x]
grid = np.linspace(start=0, stop=T, num=m)
sm_data = {'n':n, 'x':x, 'y':y, 'a':1, 'b':0.5, 'm':m, 'grid':grid}

# Initialise stan object
with open('gp_regression.stan','r',newline='') as f:
    sm = stan.build(f.read(),sm_data,random_seed=1)

# Select the number of MCMC chains and iterations, then sample
chains, samples, burn = 1, 10000, 1000
fit=sm.sample(num_chains=chains, num_samples=samples, num_warmup=burn, save_warmup=False)

# Plot regression function and posterior for rho
fig,axs=plt.subplots(1,2,figsize=(10,4),constrained_layout=True)
fig.canvas.manager.set_window_title('GP regression posterior')
f = np.mean(fit['fn_vals'], axis=1)
true_f = [reg_fn(x_i) for x_i in grid]
r = fit['rho'][0]
axs[0].plot(grid,f)
axs[0].plot(grid,true_f, color='c', lw=2, linestyle='--')
axs[0].scatter(x,y, color='black')
axs[0].set_title('Posterior mean regression function')
axs[0].set_xlabel(r'$x$')
h = axs[1].hist(r,200, density=True);
axs[1].set_title('Approximate posterior density of '+r'$\rho$')
axs[1].set_xlabel(r'$\rho$')
plt.show()
```

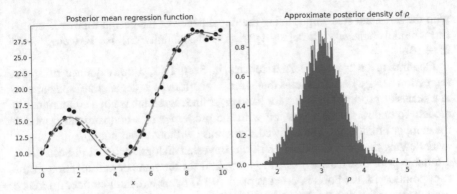

10.3 Spline Models

Linear spline regression models were introduced in Sect. 8.3.1.2 as an interesting special case of the normal linear model. For an increasing sequence of $m \geq 0$ *knot points* $\tau = (\tau_1, \ldots, \tau_m)$, the linear spline basis functions (8.16) together with some intercept terms give the following piecewise linear regression model:

$$f(x) = \alpha_0 + \alpha_1 x + \sum_{j=1}^{m} \beta_j (x - \tau_j)_+, \tag{10.10}$$

for $\alpha_j, \beta_j \in \mathbb{R}$. The regression function (10.10) is a continuous function made up of $(m + 1)$ linear segments, and can be generalised straightforwardly to continuous piecewise polynomials of degree $d \geq 1$ with $(d - 1)$ continuous derivatives:

$$f(x) = \sum_{j=0}^{d} \alpha_j x^j + \sum_{j=1}^{m} \beta_j (x - \tau_j)_+^d. \tag{10.11}$$

The special case of the regression function (10.11) where $d = 3$, corresponding to *cubic splines*, is known to present an optimal trade-off between smoothness (squared second derivative) and fidelity to fitted data points (squared residuals) (Green and Silverman 1994, see, for example, p. 11).

The general spline regression function (10.11) is a linear model with respect to the regression coefficients (*cf.* Sect. 8.3.1), and the closed-form marginal likelihood (8.12) still applies. With V now denoting the covariance matrix of the parameter vector $(\alpha_0, \ldots, \alpha_d, \beta_1, \ldots, \beta_m)$ under the conjugate prior (*cf.* Sect. 8.2.1), here the same likelihood equation is written

$$p(\mathbf{y} \mid \mathbf{x}, m, \tau) = \frac{\Gamma(a_n) \, |V_n|^{\frac{1}{2}} \, b^a}{(2\pi)^{\frac{n}{2}} \, \Gamma(a) \, |V|^{\frac{1}{2}} \, b_n^{a_n}},$$

with the left-hand side emphasising the dependency of the design matrix X of this linear model on the m-vector of knot points τ. The quantities a_n, b_n, V_n were defined in (8.10).

Continuing on from an earlier remark in Sect. 10.2, a consequence of spline regression being a linear model is that it must therefore be a (degenerate) special case of a Gaussian process. However, for fixed m, spline regression is not a nonparametric model. To endow spline regression with the properties of a nonparametric model, the number of knots must be allowed to increase without upper bound.

Following the same construction as the Bayesian histogram model in Sect. 9.4.1, a suitable prior distribution $p(m, \tau)$ is required for the number and location of the knot points, with the Poisson process prior (9.15) being a default choice. Inference from the posterior density for the knot locations,

$$p(m, \tau \mid \mathbf{y}, \mathbf{x}) \propto p(m, \tau) \, \frac{|V_n|^{\frac{1}{2}}}{|V|^{\frac{1}{2}} b_n{}^{a_n}},$$

can be achieved using (reversible jump) Markov chain Monte Carlo sampling (Green 1995) (*cf.* Chap. 5). Further, in-depth coverage of inference for nonparametric spline models is provided within (Denison et al. 2002).

Exercises 10.3 *Spline regression as a Gaussian process.* Suppose the spline regression function f from (10.11) with coefficients $(\alpha_0, \ldots, \alpha_d, \beta_1, \ldots, \beta_m) \sim$ Normal$_{m+d+1}(0, v \, I_{m+d+1})$. Express f as a Gaussian process.

10.3.1 Spline Regression with Equally Spaced Knots

Analogous to the equal bin-size histogram from Sect. 9.4.2, the spline regression inference problem can be further simplified through an assumption of equally spaced knot points on an observation interval, say $[0, T]$. Writing $p(m)$ for the prior probability mass function for the number of knots m, the conditional distribution $p(\tau \mid m)$ then assigns probability one to the m-vector τ^* with jth element

$$\tau_j^* = j \, \frac{T}{m + 1}, \quad j = 1, \ldots, m. \tag{10.12}$$

Posterior inference then concentrates on the single, unknown parameter m,

$$p(m \mid \mathbf{y}, \mathbf{x}) \propto p(m) \, \frac{|V_n|^{\frac{1}{2}}}{|V|^{\frac{1}{2}} b_n{}^{a_n}}. \tag{10.13}$$

As in Sect. 9.4.2, posterior expectations with respect to (10.13) can then be calculated directly by taking a finite sum approximation over a sufficiently large range of values for m.

Example 10.3 Consider a spline regression function with equally spaced knot points, partitioning $[0, T]$ into an unknown, geometrically distributed number of segments,

$$f(x) = \sum_{j=0}^{d} \alpha_j x^j + \sum_{j=1}^{m} \beta_j \left(x - \frac{j}{m+1} T \right)_+^d,$$

$$p(m) = (1 - \lambda)\lambda^m,$$

where $0 \le \lambda \le 1$. The following Python code (spline_regression.py) demonstrates Bayesian model averaging over the number of knot points to estimate the posterior mean regression function under the conjugate prior. The code builds upon the code linear_regression.py (see page 156), presented as a solution to Exercise 8.4 which required a marginal likelihood function (named lm_log_likelihood()) for the conjugate Bayes linear model.

For each number of knot points, spline_design_matrix() obtains the implied linear model design matrix, in order to obtain the marginal likelihood and posterior mean regression function. Following (8.8), the prior covariance matrix for the regression coefficients ($\alpha_0, \ldots, \alpha_d, \beta_1, \ldots, \beta_m$) is assumed to take the simplified form $V = \lambda^{-1} I_p$ for some scalar value $\lambda > 0$, where $p = m + d + 1$ is the number of regression coefficients.

```
## spline_regression.py

import numpy as np
from linear_regression import lm_log_likelihood

def log_prior(m,theta=.01):
    return(np.log(theta)+m*np.log(1-theta))

def spline_design_matrix(x,tau=[],d=1):
    X = np.zeros([len(x),d+1+len(tau)])
    X[:,0] = np.ones(len(x))
    for j in range(1,d+1):
        X[:,j] = [x_i**j for x_i in x]
    for j in range(len(tau)):
        for i in range(len(x)):
            if x[i] > tau[j]:
                X[i,d+1+j] = (x[i]-tau[j])**d
    return(X)

def model_average(x,y,max_m=100,grid=50,T=1,d=1,a=.1,b=1,lam=.01):
    max_post = -float('inf')
    sum_probs = 0
    ave_f = np.zeros(grid)
    probs = np.empty(max_m+1)
    x_grid = np.linspace(start=0, stop=T, num=grid)
    for m in range(max_m+1):
        tau = np.arange(1,m+1)/float(m+1)*T
        X = spline_design_matrix(x,tau,d)
        log_lhd,m_n = lm_log_likelihood(y,X,d,a,b,lam)
        log_post= log_lhd + log_prior(m)
        if log_post > max_post:
            renormalise = np.exp(max_post-log_post)
            sum_probs *= renormalise
            probs[:m] *= renormalise
            ave_f *= renormalise
            max_post = log_post
            max_m,max_m_n,max_tau = m,m_n,tau
        probs[m] = np.exp(log_post-max_post)
        sum_probs += probs[m]
        f = np.dot(spline_design_matrix(x_grid,tau,d),m_n)
        ave_f += probs[m] * f
    return(ave_f/sum_probs, max_m, max_m_n, max_tau, probs/sum_probs)
```

Next, the Python code (splines_regression_simulate.py) provides a simulated example, sampling ten noisy observations from the function (10.9) used in Sect. 10.2.2. Cubic spline regression is determined by choosing $d = 3$. Three plots are generated by the code: the true regression function (dashed line) compared against the model-averaged posterior expectation (solid line), evaluated at 50 equally spaced grid points; the approximated posterior distribution for m; the fitted spline function for the maximum a posteriori value of m, which is seen from the middle plot to be $m = 3$.

```python
#! /usr/bin/env python
## spline_regression_simulate.py
from spline_regression import model_average
import numpy as np
import matplotlib.pyplot as plt
import sys

def reg_fn(t): return(10+5*np.sin(t)+t**2/5.0)

def spline_fn(t,tau,beta,d=1):
    val = beta[0] + np.dot(beta[1:(d+1)],[t**j for j in range(1,d+1)])
    for j in range(len(tau)):
        val += beta[j+d+1]*(t-tau[j])**d if t > tau[j] else 0
    return(val)

gen = np.random.default_rng(seed=0)
n = 10 # number of observations
d = 3 if len(sys.argv)<2 else int(sys.argv[1]) # degree of splines
T = 10 # size of function domain
x = np.linspace(start=0, stop=T, num=n)
y = [gen.normal(loc=reg_fn(x_i)) for x_i in x]
grid = np.linspace(start=0, stop=T, num=50)
ave_f,max_m,max_mn,max_tau,pm=model_average(x,y,40,len(grid),T,d)
fig,axs=plt.subplots(1,3,figsize=(12,4),constrained_layout=True)
fig.canvas.manager.set_window_title('Spline regression posterior')
true_f = [reg_fn(x_i) for x_i in grid]
for plt_ind in [0,2]:
    axs[plt_ind].plot(grid,true_f, color='c', lw=2, linestyle='--')
    axs[plt_ind].scatter(x,y, color='black')
    axs[plt_ind].set_xlabel(r'$x$')
axs[0].plot(grid,ave_f)
axs[0].set_title('Posterior mean regression function')
axs[1].bar(range(len(pm)),pm)
axs[1].set_xlabel(r'$m$')
axs[1].set_title('Posterior '+r'$p(m\vert x,y)$')
axs[2].plot(grid, [spline_fn(t_i,max_tau,max_mn,d) for t_i in grid])
axs[2].set_title('Mean regression function for '+r'$m=$'+str(max_m))
plt.show()
```

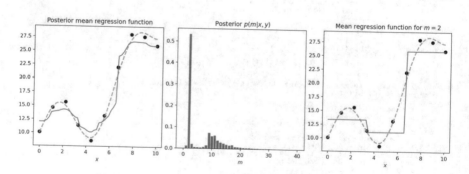

10.4   Partition Regression Models

Partition models were introduced in Sect. 9.4.1 for estimating probability distributions, demonstrating a fundamental idea that an arbitrarily complex global model can be arrived upon by adaptively partitioning the model space and assuming relatively simple statistical models within each region of the partition. This idea extends very naturally to the regression setting, suggesting an adaptive partition of the covariate space whilst assuming straightforward parametric regression models for each region of the partition. Excellent expositions of this principle are given by Holmes et al. (2005) and Denison et al. (2002, Chap. 7), the latter noting that partition models build on a premise that "points nearby in predictor space come from the same local model". Again, by assuming no upper bound to the size of the partition, partition models qualify as nonparametric models with the flexibility to approximate a broad class of regression functions.

Formally, let $\pi = \{B_1, B_2, \ldots\}$ be a partition of $\mathscr{X}$. The same parametric regression model $p(y \mid x, \theta_j)$ can be independently applied to each $\mathscr{X}$-subset B_j of the partition π, with subset-specific parameters θ_j. A simple partition model for $\mathbf{y} \mid \mathbf{x}$ thereby assumes a likelihood function of the form

$$p(\mathbf{y} \mid \mathbf{x}, \pi) = \prod_j \int_\Theta p(\mathbf{y}_j \mid \mathbf{x}_j, \theta_j) \, dQ(\theta_j), \tag{10.14}$$

where $\mathbf{x}_j$ denotes the predictor values from $\mathbf{x}$ which lie inside B_j, and $\mathbf{y}_j$ denotes the corresponding responses. More generally, a partition regression model may incorporate additional global parameters ψ, suggesting a more general likelihood function

$$p(\mathbf{y} \mid \mathbf{x}, \pi) = \int_\Psi \, dQ(\psi) \prod_j \int_\Theta p(\mathbf{y}_j \mid \mathbf{x}_j, \theta_j, \psi) \, dQ(\theta_j \mid \psi). \tag{10.15}$$

Two illustrative examples of this modelling paradigm are presented in this section: univariate changepoint models and multivariate classification and regression trees.

10.4.1 Changepoint Models

Partition models in one dimension, such that $\mathscr{X} \subseteq \mathbb{R}$ are also known as *changepoint models*. The covariate space $\mathscr{X}$ can be divided into intervals $B_j = [\tau_{j-1}, \tau_j)$ implied by an m-vector of ordered changepoints $\tau = (\tau_1, \ldots, \tau_m)$ (*cf.* Sects. 9.4.1 and 10.3). Describing a model specification as a *changepoint model* is often suggestive of some discontinuity between segments in the global regression function. In contrast, the local models within each segment will often assume exchangeability of responses, such that $p(y \mid x, \theta) = p(y \mid \theta)$. In such cases, (10.14), for example, implies

$$p(\mathbf{y} \mid \mathbf{x}, \pi) = p(\mathbf{y} \mid \mathbf{x}, m, \tau) = \prod_j p(\mathbf{y}_j).$$

If a conjugate regression model is assumed for each segment, (10.14) and (10.15) can each provide closed-form expressions for the corresponding changepoint likelihood function $p(\mathbf{y} \mid \mathbf{x}, m, \tau,)$. In such cases, assuming a suitable prior distribution $p(m, \tau)$ for the number and location of the changepoints such as the Poisson process prior (9.15), inference from the posterior density for the changepoints,

$$p(m, \tau \mid \mathbf{y}, \mathbf{x}) \propto p(m, \tau) \, p(\mathbf{y} \mid, \mathbf{x}, m, \tau),$$

can be achieved via (reversible jump) Markov chain Monte Carlo sampling (Green 1995). Python code implementing MCMC inference for some standard cases of the segment regression model $p(y \mid x, \theta_j, \psi)$ can be found in Heard (2020); this software also considers an extended modelling paradigm of changepoint *regimes*, allowing an additional complexity that regression parameters θ_j might be shared between several changepoint segments.

Exercises 10.4 (*Normal changepoint model as a Gaussian process*) Suppose $m > 0$ known changepoints $\tau = (\tau_1, \ldots, \tau_m) \in (0, T)^m$ in a piecewise constant regression function,

$$f(x) = \sum_{j=0}^{m} \mathbb{1}_{[\tau_j, \tau_{j+1})}(x) \cdot \mu_j,$$

where $\tau_0 \equiv 0$, $\tau_{m+1} \equiv T$ and independently $\mu_1, \ldots, \mu_{m+1} \sim \text{Normal}(0, v)$. Conditional on τ, express this changepoint model for f as a Gaussian process.

10.4.1.1 Changepoint Regression with Equally Spaced Changepoints

Analogously to Sects. 9.4.2 and 10.3.1, inference about an unknown number of unknown changepoint locations can be largely simplified (although possibly oversimplified) by assuming the changepoints are equally spaced on a given interval, say $[0, T]$, meaning that only the number of changepoints m is considered unknown. Writing the prior probability mass function for m as $p(m)$, and again denoting the equally spaced changepoints as τ^* (10.12), this leads to a univariate posterior mass function

$$p(m \mid \mathbf{y}, \mathbf{x}) \propto p(m) \, p(\mathbf{y} \mid \mathbf{x}, m, \tau^*). \tag{10.16}$$

Posterior inference with (10.16) can typically be well approximated analytically by taking a finite sum with a suitably large number of terms, as noted in the aforementioned sections.

Example 10.4 (Piecewise constant normal changepoint regression) Consider the equally spaced changepoint model (10.16), with exchangeable, normally distributed observations $\mathbf{y}_j$ within each segment j. Suppose the normal distribution mean parameters μ_j vary between segments, but a single, global error variance parameter σ^2 is shared by all the segments.

Assuming conjugate priors for all unknowns implies a piecewise constant regression with closed-form marginal likelihood of the form (10.15), where σ corresponds to the nuisance parameter ψ. This particular changepoint model is actually equivalent to the spline regression model in Exercise 10.3 when the degree $d = 0$, corresponding to a piecewise constant regression function.

Repeating the simulated data analysis from Exercise 10.3 with the same Python code (`spline_regression_simulate.py`) but setting $d = 0$ gives the following output plots for changepoint inference.

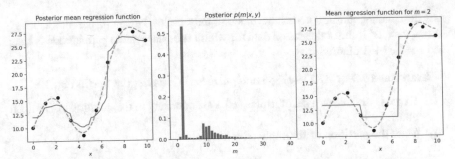

Even though the assumed regression function has m discontinuities for each value of m, the model-averaged posterior mean regression function is continuous. Conversely, because the number of observations is small ($n = 10$) the posterior mode for the number of changepoints is found at $m = 2$, despite the underlying regression function begin a smooth, non-constant function.

10.4.2 Classification and Regression Trees

Binary trees provide an interpretable class of models for recursively partitioning a multivariate predictor space $\mathcal{X} \subset \mathbb{R}^p$ with $p \geq 2$. Intuitively, they are a natural multivariate extension of univariate changepoint models. Figure 10.1 shows an illustrative tree; within each of the square terminal nodes, a separate regression model could be fit to the response data falling within that category, combining for an overall likelihood model (10.14).

Denison et al. (1998) parameterise a tree T as a set of triples of the form

$$\text{(splitting node label, variable index, splitting value).} \tag{10.17}$$

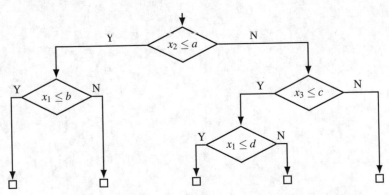

Fig. 10.1 An example of a classification and regression tree model on three variables

Any descendant splitting node label s is uniquely defined given its parent's label s', setting $s = 2s'$ if the node acts on data for which the query at the parent node is true, and $s = 2s' + 1$ otherwise.

⌑ **Exercises 10.5** (*CART notation and partition*) Consider the tree in Fig. 10.1.

(i) Express the tree as a set of triples (10.17) according to the notation of Denison et al. (1998).

(ii) State the partition of $\mathbb{R}^3$ implied by the tree.

Bayesian implementations of partition modelling with trees for classification and regression problems (CART) are described in Chipman et al. (1998) and Denison et al. (1998); each proposes a different prior distribution $p(T)$ for the partitioning tree T, and uses Markov chain Monte Carlo methods to sample from the posterior distribution of the tree. Both articles openly discuss how MCMC sampling of the trees is fraught with difficulties, due to the nested structure of the partitions. Perhaps more significantly, Chipman et al. (2010) presents a Bayesian additive regression tree (BART) model which provides further flexibility and better MCMC mixing; some Python implementations of BART can be found online.

Chapter 11
Clustering and Latent Factor Models

Hierarchical models were previously discussed in Sect. 3.3. This chapter gives further details of practical Bayesian modelling with hierarchies. In some application contexts, the hierarchies are understood to be known during the data collection process. For example, in the student-grade model of Sect. 6.1, the hierarchical structure recognised that each row of the data matrix X corresponded to test grades from the same student.

In other contexts, the hierarchies may be a subjective construct with associated uncertainty. These hierarchies are characterised by additional unknown parameters, sometimes formulated as discrete *clusters* and otherwise as continuous *latent factors*. This chapter considers some more advanced modelling techniques commonly applied in such cases.

11.1 Mixture Models

Suppose $\mathbf{x} = (x_1, \ldots, x_n)$ are n sampled continuous random variables which are assumed to be exchangeable. By De Finetti's representation theorem (Theorem 2.2), necessarily

$$p(\mathbf{x}) = \int \prod_{i=1}^{n} \mathbb{p}(x_i) \; \mathrm{d}Q(\mathbb{p}),$$

where the integral is taken over some suitable space of density functions for the unknown density $\mathbb{p}$.

A flexible class of density functions can be obtained by considering families of mixture distributions. As with the partition models considered in Sect. 9.4, each component density might be a relatively standard parametric model and yet still give rise to a mixture which is very adaptable to different underlying density shapes. The following sections present finite and infinite mixture representations, although the difference between the two can be fairly limited in practice.

The original version of this chapter has been revised due to typographic errors. The corrections to this chapter can be found at https://doi.org/10.1007/978-3-030-82808-0_12

© The Author(s), under exclusive license to Springer Nature Switzerland AG 2021, corrected publication 2022
N. Heard, *An Introduction to Bayesian Inference, Methods and Computation*, https://doi.org/10.1007/978-3-030-82808-0_11

Remark 11.1 Mixture distributions can be regarded as *clustering* models (Fraley and Raftery 2002), implicitly partitioning the n variables according to the mixture component from which they were drawn. Estimating this underlying cluster structure can sometimes be a primary inferential objective, requiring specification of a suitable loss function (Sect. 1.5.2) as exemplified by Lau and Green (2007).

11.1.1 Finite Mixture Models

Suppose the assumed density $\mathbb{p}$ is a mixture of m component densities from the same parametric family, with a general formulation

$$\mathbb{p}(x) = \sum_{j=1}^{m} w_j \, f(x \mid \theta_j, \psi), \tag{11.1}$$

where $\theta = (\theta_1, \ldots, \theta_m)$ are unknown parameters specific to each mixture component. In contrast, ψ is a global unknown parameter shared across all components, which in some settings will be redundant. The mixture weights $w = (w_1, \ldots, w_m)$ are non-negative and sum to one.

Let $\mathbf{z} = (z_1, \ldots, z_n) \in \{1, \ldots, m\}^n$ denote latent variables representing the mixture components from which each sample is drawn. Formally, $x_i \sim \mathbb{p}$ can be equivalently expressed as

$$\begin{aligned} z_i &\sim \text{Categorical}_m(w), \\ x_i &\sim f(\cdot \mid \theta_{z_i}, \psi), \end{aligned} \tag{11.2}$$

such that z_i takes value $j \in \{1, \ldots, m\}$ with probability $p(z_i = j) = w_j$, and then x_i is sampled from the z_ith-component density.

Inferring the latent variables $\mathbf{z}$ equates to clustering the observed variables $\mathbf{x}$ into at most m non-empty groups, where only samples within the same cluster are assumed to be drawn from the same population. The conditional likelihood function for $\mathbf{x}$ given the latent cluster allocations $\mathbf{z}$ is simply

$$p(\mathbf{x} \mid \mathbf{z}, \theta, \psi) = \prod_{i=1}^{n} f(x_i \mid \theta_{z_i}, \psi). \tag{11.3}$$

Marginalising the unknown parameters θ, ψ from (11.3) with respect to assumed prior distributions yields

$$p(\mathbf{x} \mid \mathbf{z}) = \int_{\Psi} \prod_{j=1}^{m} \left\{ \int_{\Theta} \prod_{i:z_i=j} f(x_i \mid \theta_j, \psi) \, dQ(\theta_j \mid \psi) \right\} \, dQ(\psi). \tag{11.4}$$

This calculation will be straightforward when assuming conjugate parametric models (*cf.* Sect. 4.2).

11.1.1.1 Dirichlet Prior for Mixture Weights

The conjugate prior for the mixture weights w is a Dirichlet distribution,

$$w \sim \text{Dirichlet}_m(\alpha_1, \ldots, \alpha_m), \tag{11.5}$$

for non-negative hyperparameters $\alpha_1, \ldots, \alpha_m$, chosen such that $\alpha = \sum_{j=1}^{m} \alpha_j$ represents a notional prior sample size (*cf.* Sect. 9.2.1). To obtain symmetry, the Dirichlet hyperparameters α_j are typically assumed to be identical with each $\alpha_j = \alpha/m$ for a chosen value of $\alpha > 0$.

For a given vector of cluster allocations $\mathbf{z}$ and for each $j \in \{1, \ldots, m\}$, let

$$n_j = \sum_{i=1}^{n} \mathbb{1}_{\{j\}}(z_i) \tag{11.6}$$

be the number of samples attributed to the jth cluster. Under the categorical model (11.2),

$$p(\mathbf{z} \mid w) = \prod_{j=1}^{m} w_j^{n_j}. \tag{11.7}$$

Marginalising (11.7) with respect to the Dirichlet prior (11.5) for the unknown mixture weights yields a marginal distribution for the cluster allocations,

$$p(\mathbf{z}) = \frac{\Gamma(\alpha)}{\Gamma(\alpha + n)} \prod_{j=1}^{m} \frac{\Gamma(\alpha_j + n_j)}{\Gamma(\alpha_j)}. \tag{11.8}$$

Remark 11.2 The probability distribution (11.8) is known as the *multinomial-Dirichlet* distribution.

Under the Dirichlet prior, the joint conditional distribution for $\mathbf{x}$ and $\mathbf{z}$ can be conveniently written up to proportionality as

$$p(\mathbf{x}, \mathbf{z} \mid \theta, \psi) \propto \prod_{j=1}^{m} \left\{ \Gamma(\alpha_j + n_j) \prod_{i:z_i=j} f(x_i \mid \theta_j, \psi) \right\}.$$

Alternatively, by first marginalising the unknown parameters (θ, ψ), the expression (11.8) can be combined with (11.4) to yield

$$p(\mathbf{x}, \mathbf{z}) \propto p(\mathbf{z}) p(\mathbf{x} \mid \mathbf{z}) \propto \int_{\Psi} \prod_{j=1}^{m} \left\{ \Gamma(\alpha_j + n_j) \int_{\Theta} \prod_{i:z_i=j} f(x_i \mid \theta_j, \psi) \, dQ(\theta_j \mid \psi) \right\} \, dQ(\psi).$$

(11.9)

11.1.1.2 Mixture of Gaussians

For densities which require support over the whole real line, f in (11.1) is commonly assumed to be the density of a normal distribution with parameter pair $\theta_j = (\mu_j, \sigma_j)$ denoting the mean and standard deviation, respectively, for the jth mixture component, implying

$$\mathbb{p}(x) = \sum_{j=1}^{m} w_j \, \phi\{(x - \mu_j)/\sigma_j\}/\sigma_j,$$

where ϕ is the standard normal density.

Assuming conjugate normal and inverse-gamma priors for $\{(\mu_j, \sigma_j) \mid j = 1, \ldots, m\}$,

$$\mu_j \mid \sigma_j \sim \text{Normal}_p(0, \sigma_j^2 \lambda^{-1}),$$
$$\sigma_j^{-2} \sim \text{Gamma}(a, b),$$

with $a, b, \lambda > 0$, the parameters μ_j and σ_j^2 can be integrated out according to (11.9) to obtain the joint distribution (11.9) of $\mathbf{x}$ and $\mathbf{z}$,

$$p(\mathbf{x}, \mathbf{z}) = \frac{\Gamma(\alpha)}{\Gamma(\alpha + n)(2\pi)^{\frac{n}{2}}} \prod_{j=1}^{m} \frac{\Gamma(\alpha_j + n_j)\Gamma(a + n_j/2) \, \lambda^{\frac{1}{2}} \, b^a}{\Gamma(\alpha_j)\Gamma(a)(\lambda + n_j)^{\frac{1}{2}} \left[b + \frac{1}{2}\{\ddot{x}_j - \dot{x}_j^2/(\lambda + n_j)\} \right]^{a + \frac{n_j}{2}}},$$

(11.10)

where $\dot{x}_j = \sum_{i:z_i=j} x_i$ and $\ddot{x}_j = \sum_{i:z_i=j} x_i^2$.

A simpler but less flexible implementation of the mixture of Gaussians model can be obtained by assuming a single variance parameter which is common to each mixture component density, such that $\theta_j = \mu_j$ and $\psi = \sigma$. The corresponding joint distribution for $\mathbf{x}$ and $\mathbf{z}$ is

$$p(\mathbf{x}, \mathbf{z}) = \frac{\Gamma(\alpha)\Gamma(a + n/2) \, \lambda^{\frac{1}{2}} \, b^a}{\Gamma(\alpha + n)\Gamma(a)(2\pi)^{\frac{n}{2}}(\lambda + n)^{\frac{1}{2}} \left[b + \frac{1}{2}\left\{ \ddot{x} - \sum_{j=1}^{m} \dot{x}_j^2/(\lambda + n_j) \right\} \right]^{a + \frac{n_j}{2}}} \prod_{j=1}^{m} \frac{\Gamma(\alpha_j + n_j)}{\Gamma(\alpha_j)},$$

where $\ddot{x} = \sum_{i=1}^{n} x_i^2$.

11.1.1.3 Inferring the Clustering and Number of Clusters

For a fixed number of mixture components m, the posterior distribution of the cluster allocations $\mathbf{z} \in \{1, \ldots, m\}^n$ can be obtained up to proportionality from the joint distribution (11.9),

$$p(\mathbf{z} \mid \mathbf{x}) \propto p(\mathbf{x}, \mathbf{z}). \tag{11.11}$$

The posterior distribution (11.11) can be explored using straightforward Markov chain Monte Carlo simulation techniques, such as Gibbs sampling, introduced in Sect. 5.3.

Exercises 11.1 (*Mixture of normals full conditionals*) For the finite mixture of normal density model (11.10) with component-specific mean and variance parameters assuming conjugate priors, state an equation, up to proportionality, for the full conditional distribution $p(z_i \mid \mathbf{z}_{-i}, \mathbf{x})$ for $i \in \{1, \ldots, n\}$.

⬛ **Exercises 11.2** (*Gibbs sampling mixture of normals*) Write code to implement Gibbs sampling on the finite mixture of normal density model (11.10) with component-specific mean and variance parameters assuming conjugate priors. Initialise the Markov chain by ordering the samples and dividing them into m equal-sized groups.

Run the code with 10,000 sampled data points generated from the mixture of two beta distributions simulated in Sect. 9.4.2.1, assuming $m = 2$. After $M = 100$ iterations, show the proportion of data points assigned to each cluster and the corresponding sample means.

More commonly the number of mixture components m will be considered unknown, requiring specification of an additional prior distribution component $p(m)$; the corresponding posterior distribution of interest extends to

$$p(m, \mathbf{z} \mid \mathbf{x}) \propto p(m) \, p(\mathbf{x}, \mathbf{z}).$$

In particular, if $p(m)$ is assumed to have unbounded support on the natural numbers (for example, assuming $m \sim \text{Poisson}(\lambda)$ for $\lambda > 0$), the finite mixture model (11.1) becomes a *potentially infinite* mixture model, and can therefore be regarded as another nonparametric inferential model akin to those considered in this chapter. Richardson and Green (1997) demonstrated inference for mixture distributions, such as mixtures of Gaussians, with an unknown number of components using reversible jump Markov chain Monte Carlo.

Remark 11.3 As with other nonparametric models, admitting an unbounded number of mixture components allows the finite mixture model (11.1) to fit increasingly complex density functions as the number of samples n increases.

11.1.2   Dirichlet Process Mixture Models

A natural evolution from the potentially infinite mixture model considered in the previous section is to consider infinite mixture models. In contrast to (11.1), suppose

$$\mathbb{p}(x) = \sum_{j=1}^{\infty} w_j \, f(x \mid \theta_j, \psi) \tag{11.12}$$

for an infinite sequence of positive-valued mixture weights $w_1, w_2, \ldots$ summing to 1, and corresponding mixture component density parameters $\theta_1, \theta_2, \ldots$. A convenient nonparametric model for obtaining infinite mixtures of type (11.12) is the Dirichlet process, introduced in Sect. 9.2.

Definition 11.1 (*Dirichlet process mixture*) The *Dirichlet process mixture* (DPM) model for $\mathbf{x} = (x_1, \ldots, x_n)$ assumes a sampling procedure where each sample x_i is drawn independently from the assumed parametric model f with a sample-specific parameter θ_i. Furthermore, each parameter θ_i is drawn independently from an unknown discrete distribution with Dirichlet process prior:

$$x_i \mid \theta_i \sim f(\cdot \mid \theta_i, \psi), \quad i = 1, \ldots, n,$$
$$\theta_i \sim G, \quad i = 1, \ldots, n,$$
$$G \sim \mathrm{DP}(\alpha \cdot \mathbb{P}_0),$$

for $\alpha > 0$ and some base probability distribution $\mathbb{P}_0$. More formally,

$$p(\mathbf{x}) = \int_{\Psi} \int_{G} \prod_{i=1}^{n} \left\{ \int_{\Theta} f(x_i \mid, \theta_i, \psi) \, \mathrm{d}G(\theta_i) \right\} \, \mathrm{d}Q(G) \, \mathrm{d}Q(\psi), \tag{11.13}$$

where $Q(G)$ is a Dirichlet process.

Remark 11.4 To ease inference, the base probability distribution $\mathbb{P}_0$ is typically assumed to be continuous and conjugate to the parametric density f.

Proposition 11.1 *Samples* $\mathbf{x} = (x_1, \ldots, x_n)$ *drawn from a DPM are exchangeable.*

Proof This property propagates automatically from the exchangeability of $\theta_1, \theta_2, \ldots$ from a distribution following a Dirichlet process (see Exercise 9.1).

Remark 11.5 It was noted in Sect. 9.2 that distributions sampled from a Dirichlet process are discrete with probability 1, and it is this discreteness which makes (11.13) a clustering model: the parameter θ_i for a sample value x_i has positive probability of matching the parameter of other samples in $\mathbf{x}$, and consequently clusters of $\mathbf{x}$ can be defined by equivalence classes of samples with the same parameter value. Similarly, because samples from a Dirichlet process with continuous base measure have countably infinite support, the model implies an infinite number of clusters.

The representation of (11.13) as a countably infinite mixture model (11.12) can be directly obtained using the *stick-breaking* interpretation of the Dirichlet process, presented in Proposition 9.1. Immediately from (9.5), the DPM model (11.13) implies a random probability density function of the form (11.12) where the atoms $\theta_1, \theta_2, \ldots$ are drawn from $\mathbb{P}_0$ and the mixture weights are determined by (9.6).

11.1.2.1   Inferring Clusters

For the infinite mixture model (11.12), the latent cluster allocation variables $\mathbf{z} = (z_1, \ldots, z_n)$ could naturally assume an infinite range of values $\{1, 2, \ldots\}$ with no upper bound. However, the labels assigned to clusters are arbitrary and at most n clusters can be non-empty. Instead, $\mathbf{z}$ can be more useful defined by revealing the samples sequentially according to the predictive distribution (9.9). Let $z_1 = 1$ and

$$z_i \in \{1, \ldots, z_{i-1}, z_{i-1} + 1\}, \quad i > 1,$$

where setting $z_i = z_{i-1} + 1$ corresponds to forming a new cluster, drawing a new parameter θ_{z_i} from the base measure $\mathbb{P}_0$; otherwise, setting $z_i = j$ for $j \in \{1, \ldots, z_{i-1},\}$ corresponds to reuse of an already drawn parameter θ_j. Further, let

$$p(z_i = j) = \begin{cases} \frac{\alpha}{\alpha+i} & \text{if } j = z_{i-1} + 1, \\ \frac{n_{j,i}}{\alpha+i} & \text{if } 1 \leq j \leq z_{i-1}, \end{cases} \tag{11.14}$$

where $n_{j,i} = \sum_{\ell=1}^{i-1} \mathbb{1}_{\{j\}}(z_\ell)$ is the number of samples allocated to cluster j prior to sample i. Then assuming (11.14) corresponds exactly to Dirichlet process sampling (see (9.9) and the subsequent remark).

At the end of the sequence, let n_j be the number of samples allocated to cluster j (11.6), and

$$m(\mathbf{z}) = \sum_{j=1}^{\infty} \mathbb{1}(n_j > 0)$$

denote the number of non-empty clusters formed. Combining the terms from (11.14), the DPM induces a marginal prior distribution for $\mathbf{z}$,

$$p(\mathbf{z}) = \frac{\Gamma(\alpha)}{\Gamma(\alpha+n)} \prod_{j=1}^{m(\mathbf{z})} \alpha \Gamma(n_j). \tag{11.15}$$

Remark 11.6 The sequential consideration of the samples used to derive (11.15) does not contradict the exchangeability property from Proposition 11.1, since (11.15) is symmetric in the indices of $\mathbf{z}$.

Remark 11.7 The Dirichlet process mixture marginal distribution for $\mathbf{z}$ (11.15) is actually very similar to the multinomial-Dirichlet prior (11.8). Although assuming an infinite number of clusters is mathematically elegant, there is little practical difference between assuming infinitely many clusters and assuming an unbounded but finite number of clusters; since when inferring cluster assignments $\mathbf{z}$, the former specification simply guarantees an infinite number of empty clusters.

Inference for the DPM is analogous to that for the multinomial Dirichlet model. The joint distribution for $(\mathbf{x}, \mathbf{z})$ can be obtained analogously to (11.9),

$$p(\mathbf{x}, \mathbf{z}) = p(\mathbf{z}) p(\mathbf{x} \mid \mathbf{z}) \propto \int_{\Psi} \prod_{j=1}^{m(\mathbf{z})} \left\{ \alpha \Gamma(n_j) \int_{\Theta} \prod_{i:z_i=j} f(x_i \mid \theta_j, \psi) \, dQ(\theta_j) \right\} \, dQ(\psi),$$

which has closed-form expression for conjugate parametric models. Posterior inference for the distribution $p(\mathbf{z} \mid \mathbf{x}) \propto p(\mathbf{x}, \mathbf{z})$ again requires Markov chain Monte Carlo simulation techniques.

11.2 Mixed-Membership Models

Consider a hierarchical random sample with two assumed layers of exchangeability, as previously considered in Example 3.2. For full generality here, first suppose $\mathbf{x} = (x_1, \ldots, x_n)$ is a vector of exchangeable random variables, each of varying dimension such that $x_i = (x_{i,1}, \ldots, x_{i,p_i})$, $p_i \geq 1$. Second, conditional on p_i, suppose the ith sample values $x_{i,1}, \ldots, x_{i,p_i}$ are also exchangeable. In this multivariate setting, *mixed-membership* clustering models can extend the mixture modelling frameworks from Sect. 11.1, by assuming a fixed but unknown distribution over mixture components *for each* sample x_i.

Formally, for a finite mixed-membership model formulation, $\mathbf{x}$ has an assumed random density function

$$p(\mathbf{x} \mid p_1, \ldots, p_n) = \prod_{i=1}^{n} \prod_{\ell=1}^{p_i} p_i(x_{i,\ell}),$$

$$p_i(x_{i,\ell}) = \sum_{j=1}^{m} w_{i,j} \, f(x_{i,\ell} \mid \theta_j, \psi), \tag{11.16}$$

where $W = (w_{i,j})$ is an $n \times m$ non-negative matrix with row sums equal to 1,

$$W \cdot \mathbf{1}_m = \mathbf{1}_n.$$

Remark 11.8 There are two points to note about the mixed-membership model (11.16).

Fig. 11.1 Belief network for
a mixed-membership model

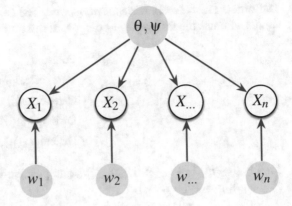

1. The mixture component densities $f(\cdot \mid \theta_j, \psi), j = 1, \ldots, n$ are common to each
   of the samples.
2. Each sample x_i has specific mixture weights $w_i = (w_{i,1}, \ldots, w_{i,m})$ for the distribution of its components $x_{i\ell}, 1 \leq \ell \leq p_i$.

Item 1 allows the learning of shared underlying populations between the samples
$x_1, \ldots, x_n$, whilst Item 2 allows for different population representations in each
sample.

The mixed-membership model (11.16) is illustrated schematically as a belief
network in Fig. 11.1.

Figure 11.1 is structurally identical to the belief network diagram for regression
modelling in Fig. 3.7. The key difference is that the shaded nodes for the mixture
weights $w_1, \ldots, w_n$ indicate that these quantities are unknown, in contrast to the measurable covariates (or *factors*) in a standard regression model; the mixture weights
can therefore be regarded as *latent factors*.

11.2.1 *Latent Dirichlet Allocation*

Mixed-membership models are frequently encountered in statistical analyses of
textual data for determining similarity amongst a collection of documents. There,
each sample x_i corresponds to a particular *document* of length p_i, such that
$(x_{i,1}, \ldots, x_{i,p_i}) \in V^{p_i}$ is the sequence of *words* in that document, which are drawn
from an overall *vocabulary* V. (Without loss of generality, it is convenient to abstract
the language of the documents by setting $V = \{1, \ldots, |V|\}$.)

The most popular mixed-membership model for text analysis is the latent Dirichlet
allocation model for categorical data, proposed by Blei et al. (2003). This finite
mixture model assumes a latent allocation variable $\mathbf{z}$ of the same dimensions as $\mathbf{x}$,
such that $z_{i,\ell} \in \{1, \ldots, m\}$ attributes a mixture component to word ℓ in document i,
$1 \leq i \leq n$ and $1 \leq \ell \leq p_i$.

Definition 11.2 (*Latent Dirichlet allocation*) The *Latent Dirichlet allocation* (LDA) model assumes the following m-component mixture model sampling procedure:

$$p_i \sim \text{Poisson}(\zeta),$$
$$x_{i,\ell} \mid z_{i,\ell}, \theta \sim \text{Categorical}_{|V|}(\theta_{z_{i,\ell}}),$$
$$z_{i,\ell} \mid w_i \sim \text{Categorical}_m(w_i),$$
$$\theta_j \sim \text{Dirichlet}_{|V|}(\alpha),$$
$$w_i \sim \text{Dirichlet}_m(\gamma),$$

for $1 \leq j \leq m$, $1 \leq i \leq n$, $1 \leq \ell \leq p_i$, and positive-valued parameters ζ, $\alpha = (\alpha_1, \ldots, \alpha_{|V|})$, $\gamma = (\gamma_1, \ldots, \gamma_m)$.

Remark 11.9 LDA can be regarded as a *latent linear model* for factorising a matrix of multinomial probabilities. Suppose $A = (a_{i,v})$ is an $n \times |V|$ matrix such that $a_{i,v}$ is the probability that a randomly selected, exchangeable element (word) of x_i is equal to $v \in V$; each row i of the matrix A therefore corresponds to a vector of multinomial word probabilities for sample i. Taking W as the $n \times m$ matrix of weights $W = (w_{i,j})$ defined above and writing $\theta = (\theta_{j,v})$ as an $(m \times |V|)$ matrix, then under the LDA model of Definition 11.2,

$$A = W \cdot \theta,$$

where the rows of both W and θ are all Dirichlet-distributed random vectors.

11.2.1.1   Topic Modelling

The LDA model in Definition 11.2 is often referred to as *topic modelling*. Under this interpretation, the m components of the mixture distribution (11.16) represent latent *topics*. Each of the m topics is characterised by a specific probability distribution on the vocabulary of words, parameterised by θ_j for topic j.

Similarly, each document is characterised by a specific mixture of the topics, parameterised by the weights w_i. The model assumes two levels of exchangeability: first amongst documents and second amongst words within a document. The latter assumption is often referred to as a *bag-of-words* model, as the ordering of words within a document is deemed unimportant by the model.

11.2.1.2   Inference

The following Stan code (`lda.stan`) is based on the example for making inference on the LDA model provided in the User's Guide of the Stan documentation,[1] adapted to the notation of this text. Stan does not support ragged array

---

[1] https://mc-stan.org/users/documentation.

data formats, and so the documents are concatenated into a single vector x; consequently, an additional variable `doc` is used to store the starting index of each document within x. The two probability vectors w_i and θ_j use the convenient `simplex` variable constraint.

```
// lda.stan

data {
  int<lower=2> m;
  int<lower=2> V;
  int<lower=1> n;
  int<lower=1> pdot;                        // sum of p_i
  int<lower=1,upper=V> x[pdot];
  int<lower=1,upper=n> doc[pdot];  // doc ID for word l
  vector<lower=0>[m] alpha;
  vector<lower=0>[V] gamma;
}
parameters {
  simplex[m] w[n];              // topic dist for document i
  simplex[V] theta[m];             // word dist for topic j
}
model {
  for (i in 1:n)
    w[i] ~ dirichlet(gamma);  // prior
  for (j in 1:m)
    theta[j] ~ dirichlet(alpha);      // prior
  for (l in 1:pdot) {
    real A[m];
    for (j in 1:m)
      A[j] = log(w[doc[l], j]) + log(theta[j, x[l]]);
    target += log_sum_exp(A);  // likelihood;
  }
}
```

11.2.2 Hierarchical Dirichlet Processes

In Sect. 11.1, Dirichlet process mixture models (DPMs, Sect. 11.1.2) were presented as an infinite-dimensional extension of finite mixture models (Sect. 11.1.1). Similarly, the finite mixed-membership model (11.16) can also naturally extend to an infinite mixture using a hierarchy of Dirichlet processes.

Definition 11.3 (*Hierarchical Dirichlet processes*) Let $\mathbb{P}_0$ be a known probability measure and $\mathbb{P}_1, \ldots, \mathbb{P}_n$ be an exchangeable sequence of unknown probability measures. Then for concentration parameters $\alpha, \gamma > 0$, a *hierarchical Dirichlet process* (Teh et al. 2006) model for $\mathbb{P}_1, \ldots, \mathbb{P}_n$, here denoted $\mathrm{HDP}(\alpha, \gamma, \mathbb{P}_0)$, assumes

$$\mathbb{P}_i \sim \mathrm{DP}(\alpha \cdot \mathbb{P}), \quad i = 1, \ldots, n,$$
$$\mathbb{P} \sim \mathrm{DP}(\gamma \cdot \mathbb{P}_0).$$

Remark 11.10 Under the hierarchical Dirichlet process, each unknown measure $\mathbb{P}_i$ has expected value $\mathbb{P}_0$.

Remark 11.11 The hierarchy introduces an additional unknown (latent) probability measure $\mathbb{P}$ which encapsulates similarities between $\mathbb{P}_1, \ldots, \mathbb{P}_n$. As $\gamma \to \infty$, the HDP model approaches n independent $\mathrm{DP}(\alpha \cdot \mathbb{P}_0)$ draws; as $\alpha \to \infty$, the n unknown probability models tend towards a single draw from $\mathrm{DP}(\gamma \cdot \mathbb{P}_0)$.

Definition 11.4 (*Hierarchical Dirichlet processes mixture*) A hierarchical Dirichlet processes mixture (HDPM) model assumes a sampling procedure where each sample component $x_{i,\ell}$ is drawn independently from a parametric model f with sample-specific parameters $\theta_{i,\ell}$ drawn independently from unknown discrete distributions:

$$x_{i,\ell} \mid \theta_{i,\ell} \sim f(\cdot \mid \theta_{i,\ell}, \psi), \quad i = 1, \ldots, n; \quad \ell = 1, \ldots, p_i,$$
$$\theta_{i,\ell} \sim G_i, \quad i = 1, \ldots, n; \quad \ell = 1, \ldots, p_i,$$
$$G_1, \ldots, G_n \sim \mathrm{HDP}(\alpha, \gamma, \mathbb{P}_0),$$

for $\alpha, \gamma > 0$ and some base probability distribution $\mathbb{P}_0$.

Proposition 11.2 *The HDPM corresponds to an infinite mixed-membership model constructed by stick-breaking process (9.4) representations for the probability density functions,*

$$\mathbb{P}_i(x) = \sum_{j=1}^{\infty} w_{i,j} \, f(x \mid \theta_j, \psi), \quad i = 1, \ldots, n, \tag{11.17}$$

where $\theta_1, \theta_2, \ldots$ are draws from the base measure $\mathbb{P}_0$ and

$$w_{i,j} = w'_{i,j} \prod_{\ell=1}^{j-1} (1 - w'_{i,\ell}),$$
$$w'_{i,j} \sim \mathrm{Beta}(\gamma \beta_j, \alpha), \quad i = 1, \ldots, n; \quad j = 1, 2, \ldots,$$
$$\beta_1, \beta_2, \ldots \sim \mathrm{GEM}(\alpha).$$

Proof See Teh et al. (2006).

11.2.2.1   Topic Modelling

The hierarchical Dirichlet process (11.17) is a mixed-membership model with an infinite number of mixture components, in contrast to the finite mixture assumed in latent Dirichlet allocation. The HDPM can be applied to topic modelling (*cf.* Sect. 11.2.1.1) on a vocabulary $V = \{1, \ldots, |V|\}$ with the following model specification:

1. $\mathbb{P}_0$ should be a $|V|$-dimensional Dirichlet distribution, such that draws $\theta_1, \theta_2, \ldots$ from $\mathbb{P}_0$ are *topics* (probability distributions on the vocabulary); then the topic distribution G_i for each document i has different atoms of mass (*topic weights*) located at the same infinite list of candidate topics.
2. Word ℓ in document i has a $\mathrm{Categorical}_{|V|}(\theta_{i,\ell})$ distribution, where $\theta_{i,\ell}$ is an independent draw from the topic distribution G_i specific to document i. Following the stick-breaking construction, in (11.17), this corresponds to

$$f(x \mid \theta_j, \psi) = \theta_{j,x}.$$

11.2.2.2 Inference

Inference for HDPM has added complexity over LDA due to the unlimited number of topics. However, open-source software implementations are available, such as the Python package *Gensim*.[2] This package uses online variational inference as described in Wang et al. (2011).

11.3 Latent Factor Models

Suppose $X = (x_{ij}) \in \mathbb{R}^{n \times p}$ is an $(n \times p)$ matrix of random variables, such that the rows of X, denoted $x_1, x_2, \ldots, x_n$, are assumed to be exchangeable p-vectors. On some occasions, particularly when the dimension $p > 1$ may be large, it might be believed that the vectors x_i lie close to a lower dimensional subspace of $\mathbb{R}^p$. In this case, probabilistic beliefs about X may be more easily characterised by specifying probability distributions in the lower dimensional space. One approach for modelling in alternative dimensions is to deploy *latent factor models*.

The canonical example of latent factor modelling assumes the following latent linear model (Bhattacharya and Dunson 2011):

$$x_i = \Lambda \cdot \eta_i + \epsilon_i, \quad i = 1, \ldots, n, \tag{11.18}$$

where

$$\epsilon_i \sim \text{Normal}_p(0, \Sigma),$$
$$\eta_i \sim \text{Normal}_k(0, I_k). \tag{11.19}$$

The elements of the vector $\eta_i \in \mathbb{R}^k$ are referred to as the *latent factors* for sample i. Typically, in dimension-reduction applications, the latent dimension $k \ll p$. The global parameter Λ is a $(p \times k)$ matrix of *factor loadings* which project the latent factors into the higher dimensional space $\mathbb{R}^p$. As η_i varies over $\mathbb{R}^k$, $\Lambda \cdot \eta_i$ defines a linear subspace of $\mathbb{R}^p$, but the observable variables x_i lie just outside that subspace due to the observation error ϵ_i.

Remark 11.12 For each sample, the latent factors $\eta_i \in \mathbb{R}^k$ can be interpreted as the unobserved measurements of k *features* which are believed to be linearly related to the expected value of the response.

Since (11.18) is a linear model, assuming (11.19) implies the latent factors can be marginalised out similarly to (8.11), yielding

$$x_i \mid \Lambda, \Sigma \sim \text{Normal}_p(0, \Lambda\Lambda^\mathsf{T} + \Sigma). \tag{11.20}$$

---

[2] https://radimrehurek.com/gensim/models/hdpmodel.html.

Remark 11.13 The marginal distribution (11.20) gives insight into the latent factor model; the Gram matrix $\Lambda\Lambda^\mathsf{T}$ of the rows of the latent factor loadings Λ provides a low-rank ($k < p$) additive contribution to the covariance matrix for each exchangeable data row x_i.

For any semi-orthogonal matrix U satisfying $UU^\mathsf{T} = I_k$, $(\Lambda U) \cdot (\Lambda U)^\mathsf{T} = \Lambda\Lambda^\mathsf{T}$, and so the covariance factorisation in (11.20) is not unique. In determining a prior distribution for Λ, it is therefore natural to choose a distribution which is invariant to these rotations and reflections, satisfying

$$p(\Lambda) = p(\Lambda U)$$

for any semi-orthogonal matrix U.

11.3.1 Stan Implementation

The following Stan code (`latent_factors.stan`) implements the latent factor model from (11.20). For simplicity, Σ is assumed to be a diagonal matrix of independent inverse-gamma distributed random variables, and a reference prior is assumed for the factor loadings.

```
// latent_factors.stan

data {
    int<lower=0> n;  // number of observations
    int<lower=1> p;  // number of grid points
    vector[p] X[n];  // data matrix
    int<lower=1> k;  // number of latent factors
    real<lower=0> a;
    real<lower=0> b;
}
parameters {
    vector<lower=0>[p] Sigma;  // diagonal Sigma
    matrix[p, k] Lambda;  // factor loadings
}
transformed parameters{
    matrix[p, p] Omega = Lambda * Lambda' + diag_matrix(Sigma);
}
model {
    Sigma ~ inv_gamma(a,b);
    X ~ multi_normal(rep_vector(0, p), Omega);
}
```

To illustrate inference for the latent factor model using `latent_factors.stan`, the following PyStan code (`latent_factors_stan.py`) simulates a (50×8) data matrix X from the model and performs posterior inference on Λ and Σ.

```python
#! /usr/bin/env python
## latent_factors_stan.py

import stan
import numpy as np
import matplotlib.pyplot as plt
from scipy.linalg import orthogonal_procrustes

# Simulate data
gen = np.random.default_rng(seed=0)
n = 50
p = 8
k = 3
Lambda = 10 * np.reshape(gen.normal(size=p*k),[p,k])
Sigma = 1.0/gen.gamma(1,1,p)
Omega = Lambda.dot(Lambda.T) + np.diag(Sigma)
X = gen.multivariate_normal(np.zeros(p),Omega,size=n)
sm_data = {'n':n, 'p':p, 'X':X, 'k':k, 'a':1, 'b':1}

# Initialise stan object
with open('latent_factors.stan','r',newline='') as f:
    sm = stan.build(f.read(),sm_data,random_seed=1)

# Select the number of MCMC chains and iterations, then sample
chains, samples, burn = 1, 10000, 1000
fit=sm.sample(num_chains=chains, num_samples=samples, num_warmup=burn, save_warmup=False)

# Perform Procrustes alignment of sampled Lambdas
lam_hat = fit['Lambda'][:,:,-1]
for i in range(samples-1):
    l = fit['Lambda'][:,:,i]
    R = orthogonal_procrustes(l,fit['Lambda'][:,:,-1])[0]
    lam_hat += l.dot(R)

lam_hat /= samples
lam_hat = lam_hat.dot(orthogonal_procrustes(lam_hat,Lambda)[0])

lam_bar = np.mean(fit['Lambda'],axis=2)
lam_bar = lam_bar.dot(orthogonal_procrustes(lam_bar,Lambda)[0])

# Plot estimate and true values for Lambda
fig,axs=plt.subplots(1,3,figsize=(7,4),constrained_layout=True)
fig.canvas.manager.set_window_title('Latent factor lambdas')
axs[0].imshow(lam_hat, cmap='Blues')
axs[0].set_title(r'$\hat{\Lambda}$')
axs[1].imshow(Lambda, cmap='Blues')
axs[1].set_title(r'$\Lambda$')
axs[2].imshow(lam_bar, cmap='Blues')
axs[2].set_title('Crude estimate '+r'$\bar{\Lambda}$')
out=plt.setp(plt.gcf().get_axes(), xticks=[], yticks=[]);
plt.show()
```

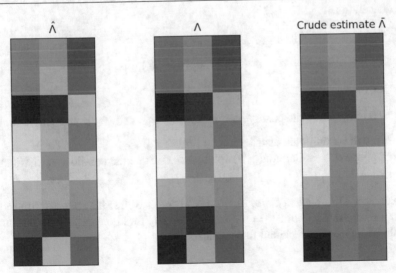

To see how well the underlying latent structure is recovered, the posterior distribution for Λ is compared with the true value used to generate X. This comparison
is not completely straightforward, since it was noted above that the model (11.20) is

invariant to semi-orthogonal transformations. This invariance to certain transformations, such as rotations of Λ, implies taking a simple average of the posterior samples
of Λ would not give a meaningful estimate.

To enable posterior averaging, the MCMC samples are first aligned using *Pro-
crustes alignment*, as advocated in Oh and Raftery (2007). Each sample is transformed by a different semi-orthogonal matrix optimised to be as close as possible
to a fixed target, here chosen to be the final MCMC sample; the aligned samples
are then averaged to obtain a posterior mean value, and finally this posterior mean
is transformed in order to be aligned as closely as possible to the true value Λ. The
resulting estimate from this post-processing procedure is here denoted $\hat{\Lambda}$.

Heat map plots of $\hat{\Lambda}$ and the true value are compared side by side in the plot
generated by the code. To demonstrate the value of this alignment procedure, the
plots also show the crude estimate, denoted $\bar{\Lambda}$, obtained from directly averaging the
posterior samples of Λ and then finding the closest alignment to the true Λ. The
estimate obtained from mutually aligning the samples is much closer to the true
matrix of factor loadings.

Exercises 11.3 (*Latent factor linear model*) Let $y = (y_1, \ldots, y_n)$ be an n-vector
of real-valued response variables, with an associated $n \times p$ matrix of covariates X
with rows $x_1, \ldots, x_n \in \mathbb{R}^p$. Consider the latent factor linear model,

$$y_i = x_i \cdot \beta + z_i \cdot \gamma + \epsilon_i,$$

which presumes an $n \times q$ matrix Z of further, unobserved covariates $z, \ldots, z_n \in \mathbb{R}^q$
with corresponding regression coefficients $\gamma \in \mathbb{R}^q$. Suppose the following independent distributions:

$$\epsilon_i \sim \text{Normal}(0, \sigma^2),$$
$$\beta \sim \text{Normal}_p(0, \sigma^2 V),$$
$$\gamma \sim \text{Normal}_q(0, \sigma^2 U),$$

for $\sigma > 0$ and symmetric, positive semidefinite $p \times p$ and $q \times q$ matrices V and U.

(i) State the conditional distribution $[y \mid \sigma, X, Z]$.
(ii) Suppose $\sigma^{-2} \sim \text{Gamma}(a, b)$ for $a, b > 0$. State the conditional distribution
    $[y \mid X, Z]$.

⌨ **Exercises 11.4** (*Latent factor linear model code*) Write Stan code to fit the model
from Exercise 11.3 with $V = v\, I_p$ and $U = u\, I_q$ for known $v, u > 0$. Assume a
reference prior for the latent factor matrix Z.

Correction to: An Introduction to Bayesian Inference, Methods and Computation

Correction to:
N. Heard, *An Introduction to Bayesian Inference,* Methods and Computation, https://doi.org/10.1007/978-3-030-82808-0

After publication of the book, the author noticed that the typeset of 'p' was incorrectly processed in Chaps. 2 and 9 during production. Therefore corrections have been incorporated in these two chapters: the typesetting of '$\mathbb{P}$' was corrected.

The updated version of these chapters can be found at
https://doi.org/10.1007/978-3-030-82808-0,
https://doi.org/10.1007/978-3-030-82808-0_2,
https://doi.org/10.1007/978-3-030-82808-0_5,
https://doi.org/10.1007/978-3-030-82808-0_6,
https://doi.org/10.1007/978-3-030-82808-0_8,
https://doi.org/10.1007/978-3-030-82808-0_9,
https://doi.org/10.1007/978-3-030-82808-0_11

Appendix A
Conjugate Parametric Models

For each probability model below, $\mathbf{x} = (x_1, \ldots, x_n)$ are n independent samples from a likelihood distribution $p(x|\theta, \psi)$, for which there exists a conjugate prior distribution $p(\theta)$ for θ.

Each of the tables for discrete and continuous parametric models provides the following details:

- Ranges for x and θ.
- Likelihood distribution $p(x \mid \theta, \psi)$ and the conjugate prior density $p(\theta)$.
- Marginal likelihood $p(\mathbf{x})$ and the posterior density $p(\theta \mid \mathbf{x})$, denoted $\pi(\theta)$.
- Posterior predictive distribution $p(x \mid \mathbf{x})$ for a new observation x.

A.1 Notation

To make notation concise, let

$$\dot{x} = \sum_{i=1}^{n} x_i, \qquad \ddot{x} = \sum_{i=1}^{n} x_i \cdot x_i^{\mathsf{T}}, \tag{A.1}$$

respectively, denote the sum and the sum of squared values in $\mathbf{x}$. Let $x_{(1)} \leq \ldots \leq x_{(n)}$ denote the order statistics of $\mathbf{x}$. Finally, for discrete random variables on $\{1, \ldots, k\}$, let

$$n_j = \sum_{i=1}^{n} \mathbb{1}_{\{j\}}(x_i) \tag{A.2}$$

be the number of samples for which $x_i = j$, and let $\mathbf{n} = (n_1, \ldots, n_k)$.

Some other items appearing in the following tables:

- Hyperparameters a, b represent positive real numbers unless otherwise stated.

© The Editor(s) (if applicable) and The Author(s), under exclusive license to Springer Nature Switzerland AG 2021
N. Heard, *An Introduction to Bayesian Inference, Methods and Computation*,
https://doi.org/10.1007/978-3-030-82808-0

- $\zeta(a, b) = \sum_{x=0}^{\infty}(x + b)^{-a}$ is the Hurwitz zeta function.
- $\Delta(k)$ denotes the standard (or probability) simplex $\{u \in \mathbb{R}^k : u_i \geq 0, \sum_{i=1}^{k} u_i = 1\}$.
- For the Dirichlet distribution, $\alpha \in \{u \in \mathbb{R}^k : u_i \geq 0, \sum_{i=1}^{k} u_i > 0\}$.
- For the normal and inverse Wishart equations, $m \in \mathbb{R}^k$ and the matrices V, ψ and B are assumed positive definite.

A.2 Discrete Models

Uniform$(x \mid \{1, \ldots, \theta\})$	Zeta$(\theta \mid a, b)$
$x \in \{1, \ldots, \theta\}$	$\theta \in \{b, b+1, \ldots\}, \quad a > 1, b \in \{1, 2, \ldots\}$
$p(x \mid \theta) = \dfrac{\mathbb{1}_{\{1,\ldots,\theta\}}(x)}{\theta}$	$p(\theta) = \dfrac{\mathbb{1}_{\{b,b+1,\ldots\}}(\theta)}{\zeta(a, b)\theta^a}$
$p(\mathbf{x}) = \dfrac{\zeta(a + n, b_*)}{\zeta(a, b)}, \quad b_* = \max\{b, x_{(n)}\}$	$\pi(\theta) = \dfrac{\mathbb{1}_{\{b_*, b_*+1, \ldots\}}(\theta)}{\zeta(a + n, b_*)\theta^{a+n}}$
$p(x \mid \mathbf{x}) = \dfrac{\zeta(a + n + 1, \max\{b_*, x\})}{\zeta(a + n, b_*)}$	$\equiv \text{Zeta}(\theta \mid a + n, b_*)$
Bernoulli$(x \mid \theta)$	**Beta$(\theta \mid a, b)$**
$x \in \{0, 1\}$	$\theta \in [0, 1]$
$p(x \mid \theta) = \theta^x (1 - \theta)^{1-x}$	$p(\theta) = \dfrac{\Gamma(a + b)}{\Gamma(a)\,\Gamma(b)}\theta^{a-1}(1 - \theta)^{b-1}$
$p(\mathbf{x}) = \dfrac{\Gamma(a + b)\,\Gamma(a + \dot{x})\,\Gamma(b + n - \dot{x})}{\Gamma(a)\,\Gamma(b)\,\Gamma(a + b + n)}$	$\pi(\theta) = \dfrac{\Gamma(a + b + n)\theta^{a+\dot{x}-1}(1 - \theta)^{b+n-\dot{x}-1}}{\Gamma(a + \dot{x})\,\Gamma(b + n - \dot{x})}$
$p(x \mid \mathbf{x}) = \dfrac{a + \dot{x}}{a + b + n - \dot{x}}$	$\equiv \text{Beta}(a + \dot{x}, b + n - \dot{x})$
Geometric$(x \mid \theta)$	**Beta$(\theta \mid a, b)$**
$x \in \{0, 1, 2, \ldots\}$	$\theta \in [0, 1]$
$p(x \mid \theta) = \theta(1 - \theta)^x$	$p(\theta) = \dfrac{\Gamma(a + b)}{\Gamma(a)\Gamma(b)}\theta^{a-1}(1 - \theta)^{b-1}$
$p(\mathbf{x}) = \dfrac{\Gamma(a + b)\,\Gamma(a + n)\,\Gamma(b + \dot{x})}{\Gamma(a)\,\Gamma(b)\,\Gamma(a + b + n + \dot{x})}$	$\pi(\theta) = \dfrac{\Gamma(a + b + n + \dot{x})\,\theta^{a+n-1}(1 - \theta)^{b+\dot{x}-1}}{\Gamma(a + n)\,\Gamma(b + \dot{x})}$
$p(x \mid \mathbf{x}) = \dfrac{(a + n)\,\Gamma(b + \dot{x} + x)\,\Gamma(a + b + n + \dot{x})}{\Gamma(b + \dot{x})\,\Gamma(a + b + n + 1 + \dot{x} + x)}$	$\equiv \text{Beta}(a + n, b + \dot{x})$
Poisson$(x \mid \theta)$	**Gamma$(\theta \mid a, b)$**
$x \in \{0, 1, 2, \ldots\}$	$\theta \in [0, \infty)$
$p(x \mid \theta) - \dfrac{\theta^x e^{-\theta}}{x!}$	$p(\theta) = \dfrac{b^a}{\Gamma(a)}\theta^{a-1}e^{-b\theta}$
$p(\mathbf{x}) = \dfrac{\Gamma(a + \dot{x})\,b^a}{\Gamma(a)\,(b + n)^{a+\dot{x}}}$	$\pi(\theta) = \dfrac{(b + n)^{a+\dot{x}}}{\Gamma(a + \dot{x})}\theta^{a+\dot{x}-1}e^{-(b+n)\theta}$
$p(x \mid \mathbf{x}) = \dfrac{\Gamma(a + \dot{x} + x)\,(b + n)^{a+\dot{x}}}{\Gamma(a + \dot{x})\,(b + n + 1)^{a+\dot{x}+x}}$	$\equiv \text{Gamma}(a + \dot{x}, b + n)$
Multinomial$_k(x \mid 1, \theta)$	**Dirichlet$_k(\theta \mid \alpha)$**
$x \in \{(1, 0, \ldots, 0), \ldots, (0, \ldots, 0, 1)\}$	$\theta \in \Delta(k)$
$p(x \mid \theta) = \theta_x$	$p(\theta) = \dfrac{\Gamma(\sum_{j=1}^{k}\alpha_j)}{\prod_{j=1}^{k}\Gamma(\alpha_j)}\prod_{j=1}^{k}\theta_j^{\alpha_j-1}$
$p(\mathbf{x}) = \dfrac{\Gamma(\sum_{j=1}^{k}\alpha_j)}{\Gamma(\sum_{j=1}^{k}\alpha_j + n)}\prod_{j=1}^{k}\dfrac{\Gamma(\alpha_j + n_j)}{\Gamma(\alpha_j)}$	$\pi(\theta) = \dfrac{\Gamma(\sum_{j=1}^{k}\alpha_j + n)}{\prod_{j=1}^{k}\Gamma(\alpha_j + n_j)}\prod_{j=1}^{k}\theta_j^{\alpha_j+n_j-1}$
$p(x \mid \mathbf{x}) = \dfrac{\alpha_j + n_j}{\sum_{j=1}^{k}\alpha_j + n}$	$\equiv \text{Dirichlet}_k(\theta \mid \alpha + \mathbf{n})$

A.3 Continuous Models

$\text{Uniform}(x \mid 0, \theta)$	$\text{Pareto}(\theta \mid a, b)$
$x \in [0, \infty)$	$\theta \in (0, \infty)$
$p(x \mid \theta) = \dfrac{\mathbb{1}_{[0,\theta]}(x)}{\theta}$	$p(\theta) = \dfrac{ab^a \, \mathbb{1}_{[b,\infty)}(\theta)}{\theta^{a+1}}$
$p(\mathbf{x}) = \dfrac{ab^a}{(a+n)b_*{}^{a+n}}, \quad b_* = \max\{b, x_{(n)}\}$	$\pi(\theta) = \dfrac{(a+n)b_*{}^{a+n} \, \mathbb{1}_{[b_*,\infty)}(\theta)}{\theta^{a+n+1}}$
$p(x \mid \mathbf{x}) = \dfrac{(a+n)b_*{}^{a+n}}{(a+n+1)\max\{b_*, x\}^{a+n+1}}$	$\equiv \text{Pareto}(\theta \mid a+n, b_*)$
$\text{Exponential}(x \mid \theta)$	$\text{Gamma}(\theta \mid a, b)$
$x \in [0, \infty)$	$\theta \in (0, \infty)$
$p(x \mid \theta) = \theta e^{-\theta x}$	$p(\theta) = \dfrac{b^a}{\Gamma(a)}\theta^{a-1}e^{-b\theta}$
$p(\mathbf{x}) = \dfrac{\Gamma(a+n)\, b^a}{\Gamma(a)\,(b+\dot{x})^{a+n}}$	$\pi(\theta) = \dfrac{(b+\dot{x})^{a+n}}{\Gamma(a+n)}\theta^{a+n-1}\,e^{-(b+\dot{x})\theta}$
$p(x \mid \mathbf{x}) = \dfrac{(a+n)(b+\dot{x})^{a+n}}{(b+\dot{x}+x)^{a+n+1}}$	$\equiv \text{Gamma}(\theta \mid a+n, b+\dot{x})$
$\text{Gamma}(x \mid \psi, \theta)$	$\text{Gamma}(\theta \mid a, b)$
$x \in [0, \infty)$	$\theta \subset (0, \infty)$
$p(x \mid \theta) = \dfrac{\theta^{\psi}}{\Gamma(\psi)}x^{\psi-1}e^{-\theta x}$	$p(\theta) = \dfrac{b^a}{\Gamma(a)}\theta^{a-1}e^{-b\theta}$
$p(\mathbf{x}) = \dfrac{\Gamma(a+n\psi)\, b^a}{\Gamma(a)\,(b+\dot{x})^{a+n\psi}}$	$\pi(\theta) = \dfrac{(b+\dot{x})^{a+n\psi}}{\Gamma(a+n\psi)}\theta^{a+n\psi-1}\,e^{-(b+\dot{x})\theta}$
$p(x \mid \mathbf{x}) = \dfrac{\Gamma(a+(n+1)\psi)\,(b+\dot{x})^{a+n\psi}}{\Gamma(a+n\psi)\,(b+\dot{x}+x)^{a+(n+1)\psi}}$	$\equiv \text{Gamma}(\theta \mid a+n\psi, b+\dot{x})$
$\text{Normal}_k(x \mid \theta, \psi)$	$\text{Normal}(\theta \mid m, V)$
$x \in \mathbb{R}^k$	$\theta \in \mathbb{R}^k$
$p(x \mid \theta) = \dfrac{\exp\{-\frac{1}{2}\sum_{i=1}^{n}(x_i-\theta)^\mathsf{T}\psi^{-1}(x_i-\theta)\}}{(2\pi)^{\frac{nk}{2}}\,|\psi|^{\frac{n}{2}}}$	$p(\theta) = \dfrac{\exp\{-\frac{1}{2}(\theta-m)^\mathsf{T}V^{-1}(\theta-m)\}}{(2\pi)^{\frac{k}{2}}\,|V|^{\frac{1}{2}}}$
$p(\mathbf{x}) = \dfrac{|V_*|^{\frac{1}{2}}\exp\{-\frac{1}{2}m^\mathsf{T}V^{-1}m - \frac{1}{2}\sum_{i=1}^{n}x_i^\mathsf{T}\psi^{-1}x_i\}}{(2\pi)^{\frac{nk}{2}}\,|\psi|^{\frac{n}{2}}\,|V|^{\frac{1}{2}}\exp\{-\frac{1}{2}m_*^\mathsf{T}V_*^{-1}m_*\}}$	$\pi(\theta) = \dfrac{\exp\{-\frac{1}{2}(\theta-m_*)^\mathsf{T}V_*^{-1}(\theta-m_*)\}}{(2\pi)^{\frac{k}{2}}\,|V_*|^{\frac{1}{2}}}$
$m_* = V_*(V^{-1}m + \psi^{-1}\dot{x}),\; V_* = (V^{-1}+n\psi^{-1})^{-1}$	$\equiv \text{Normal}(\theta \mid m_*, V_*)$
$p(x \mid \mathbf{x}) = \text{Normal}(x \mid m_*, \psi + V_*)$	
$\text{Normal}_k(x \mid 0, \theta)$	$\text{Inverse Wishart}(\theta \mid a, B)$
$x \in \mathbb{R}^k$	$\theta \in \mathbb{R}^{k \times k}$, positive definite
$p(x \mid \theta) = \dfrac{\exp\{-\frac{1}{2}\sum_{i=1}^{n}x_i^\mathsf{T}\theta^{-1}x_i\}}{(2\pi)^{\frac{nk}{2}}\,|\theta|^{\frac{n}{2}}}$	$p(\theta) = \dfrac{|B|^{\frac{a}{2}}\exp\{-\frac{1}{2}\operatorname{tr}(B\theta^{-1})\}}{2^{\frac{ak}{2}}\,|\theta|^{\frac{a+k+1}{2}}\,\pi^{\frac{k(k-1)}{4}}\,\prod_{\ell=1}^{k}\Gamma(\frac{a+1-\ell}{2})}$
$p(\mathbf{x}) = \dfrac{|B|^{\frac{a}{2}}}{|B+\ddot{x}|^{\frac{a+n}{2}}}\prod_{\ell=1}^{k}\dfrac{\Gamma(\frac{a+1-\ell}{2})}{\Gamma(\frac{a+n+1-\ell}{2})}$	$\pi(\theta) = \dfrac{|B+\ddot{x}|^{\frac{a+n}{2}}\exp\{-\frac{1}{2}\operatorname{tr}((B+\ddot{x})\theta^{-1})\}}{2^{\frac{(a+n)k}{2}}\,|\theta|^{\frac{a+n+k+1}{2}}\,\pi^{\frac{k(k-1)}{4}}\,\prod_{\ell=1}^{k}\Gamma(\frac{a+n+1-\ell}{2})}$
$p(x \mid \mathbf{x}) = \dfrac{|B+\ddot{x}|^{\frac{a+n}{2}}}{|B+\ddot{x}+x\cdot x^\mathsf{T}|^{\frac{a+n+1}{2}}}\prod_{\ell=1}^{k}\dfrac{\Gamma(\frac{a+n+1-\ell}{2})}{\Gamma(\frac{a+n+2-\ell}{2})}$	$\equiv \text{Inverse Wishart}(\theta \mid a+n, B+\ddot{x})$

Appendix B
Solutions to Exercises

Solution 1.1 *Linear transformations of utilities.* Let $u(\cdot)$ be a utility function with corresponding expected utility $\bar{u}(\cdot)$, and consider a linear transformation

$$u'(c) = \alpha + \beta\, u(c),$$

where $\alpha, \beta \in \mathbb{R}$. Under utility function $u'(\cdot)$, the corresponding expected utility for an action $a = \{(E_1, c_1), (E_2, c_2), \ldots\} \in \mathcal{A}$ is

$$\bar{u}'(a) = \sum_i \mathbb{P}(E_i)\, u'(c_i) = \alpha + \beta \sum_i \mathbb{P}(E_i)\, u(c_i) = \alpha + \beta\bar{u}(a).$$

If $\beta > 0$, then for two actions $a, a' \in \mathcal{A}$, $\bar{u}'(a) < \bar{u}'(a') \iff \bar{u}(a) < \bar{u}(a')$.

Solution 1.2 *Bounded utility.* Let $a = \{(\Omega, c)\}$ and $a' = \{(S_{u(c)}, c^*), (\overline{S_{u(c)}}, c_*)\}$. Since $\mathbb{P}(\Omega) = 1$, $\bar{u}(a) = u(c)$. For the dichotomy a', $\bar{u}(a') = \mathbb{P}(S_{u(c)})u(c^*) + (1 - \mathbb{P}(S_{u(c)}))u(c_*) = u(c).1 + (1 - u(c)).0 = u(c)$. Hence $\bar{u}(a) = \bar{u}(a')$ and therefore $a \sim a'$.

Solution 1.3 *Unbounded utility.*

   (i) If $\{\Omega, c_1)\} \sim \{(\overline{S_x}, c), (S_x, c_2)\}$, then $0 = u(c_1) = (1 - x)\, u(c) + x\, u(c_2) = (1 - x)\, u(c) + x.1 \implies u(c) = -x/(1 - x) < 0$.
   (ii) If $\{\Omega, c_2)\} \sim \{(\overline{S_x}, c_1), (S_x, c)\}$, then $1 = u(c_2) = (1 - x)\, u(c_1) + x\, u(c) = (1 - x).0 + x\, u(c) \implies u(c) = 1/x > 1$.

Solution 1.4 *Transitivity of preference.* Show that for $a, a', a'' \in \mathcal{A}$, if $a \le a'$ and $a' \le a''$, then $a \le a''$.

Solution 1.5 *Coherence with probabilities.* By Axiom 3, for any $c_1 <_C c_2, \exists\, x, x' \in [0, 1]$ such that $\{(E, c_1), (\overline{E}, c_2)\} \sim \{(S_x, c_1), (\overline{S_x}, c_2)\}$ and $\{(F, c_1), (\overline{F}, c_2)\} \sim \{(S_{x'}, c_1), (\overline{S_{x'}}, c_2)\}$, namely, $x = \mathbb{P}(E), x' = \mathbb{P}(F)$ and hence $x \le x'$. Since $x \le x'$,

© The Editor(s) (if applicable) and The Author(s), under exclusive license to Springer Nature Switzerland AG 2021
N. Heard, *An Introduction to Bayesian Inference, Methods and Computation*,
https://doi.org/10.1007/978-3-030-82808-0

$S_x \subseteq S_{x'}$, and therefore by Axiom 2 $\{(S_x, c_1), (\overline{S_x}, c_2)\} \leq \{(S_{x'}, c_1), (\overline{S_{x'}}, c_2)\}$ and the result follows from the initial equivalences and transitivity of preferences.

An alternative proof could have used the principle of maximising expected utility.

Solution 1.6 *Absolute loss (also known as L_1 loss).* For a univariate, continuous-valued $\omega \in \mathbb{R}$, the absolute loss function gives expected utility

$$\bar{u}(d_{\hat{\omega}}) = -\int_{-\infty}^{\infty} |\hat{\omega} - \omega| \, f(\omega) \, d\omega = -\int_{-\infty}^{\hat{\omega}} (\hat{\omega} - \omega) \, f(\omega) \, d\omega + \int_{\hat{\omega}}^{\infty} (\hat{\omega} - \omega) \, f(\omega) \, d\omega.$$

Differentiating the right-hand side with respect to $\hat{\omega}$ and setting equal to zero yields

$$\left\{ \int_{-\infty}^{\hat{\omega}} f(\omega) \, d\omega + \hat{\omega} \, f(\hat{\omega}) \right\} - \hat{\omega} \, f(\hat{\omega}) = \left\{ \int_{\hat{\omega}}^{\infty} f(\omega) \, d\omega - \hat{\omega} \, f(\hat{\omega}) \right\} + \hat{\omega} \, f(\hat{\omega})$$

$$\Longleftrightarrow \int_{-\infty}^{\hat{\omega}} f(\omega) \, d\omega = \int_{\hat{\omega}}^{\infty} f(\omega) \, d\omega$$

$$\Longleftrightarrow \int_{-\infty}^{\hat{\omega}} f(\omega) \, d\omega = \frac{1}{2},$$

since necessarily $\int_{-\infty}^{\infty} f(\omega) \, d\omega = 1$. Hence $\hat{\omega}$ is the median.

Solution 1.7 *Squared loss (also known as L_2 loss).*

For a univariate, continuous-valued $\omega \in \mathbb{R}$, the squared loss function gives expected utility

$$\bar{u}(d_{\hat{\omega}}) = -\int_{-\infty}^{\infty} (\hat{\omega} - \omega)^2 \, f(\omega) \, d\omega.$$

Differentiating with respect to $\hat{\omega}$ and setting equal to zero yields

$$0 = -2 \int (\hat{\omega} - \omega) \, f(\omega) \, d\omega = -2\{\hat{\omega} - \mathbb{E}(\omega)\}$$

$$\Longrightarrow \hat{\omega} = \mathbb{E}(\omega).$$

Solution 1.8 *Zero-one loss (also known as L_∞ loss).* For a univariate, continuous-valued $\omega \in \mathbb{R}$, and for $\epsilon > 0$ define the ϵ-ball zero-one loss function

$$\ell_\epsilon(\hat{\omega}, \omega) = 1 - \mathbb{1}_{B_\epsilon(\hat{\omega})}(\omega),$$

where $B_\epsilon(\omega) = (\omega - \epsilon, \omega + \epsilon)$. This loss function implies an expected utility

$$\bar{u}_\epsilon(d_{\hat{\omega}}) = \mathbb{E}[\mathbb{1}_{B_\epsilon(\hat{\omega})}(\omega)] = \mathbb{P}\{B_\epsilon(\hat{\omega})\}.$$

As $\epsilon \to 0$ to obtain the zero-one loss function, the right-hand side tends to $\epsilon f(\hat{\omega})$ which is clearly maximised by the mode of f.

Solution 1.9 *KL-divergence non-negative.* If $p = q$ then $\mathrm{KL}(p \parallel q) = \int p(x)$ log $1 \, dx = 0$.

For $p \neq q$, the non-negativity of KL-divergence can be demonstrated using the logarithmic inequality $\log(a) \geq 1 - a^{-1}$ for any $a > 0$. This rule gives

$$\log \frac{p(x)}{q(x)} \geq 1 - \frac{q(x)}{p(x)},$$

$$\implies \mathrm{KL}(p \parallel q) = \int p(x) \log \frac{p(x)}{q(x)} \, dx \geq \int p(x) \left(1 - \log \frac{q(x)}{p(x)} \right) dx = \int \{p(x) - q(x)\} \, dx = 0.$$

Therefore, when KL-divergence is used as a loss function for prediction, the smallest expected loss (zero) is incurred when reporting genuine beliefs.

Solution 2.1 *Finitely exchangeable binary sequences.* Suppose $X_1, \ldots, X_n$ are assumed to be independent and identically distributed Bernoulli($\frac{1}{2}$) random variables, and it is observed that $\sum X_i = s$. Conditional on this information, $X_1, \ldots, X_n$ are still exchangeable with constant probability mass function

$$\mathbb{P}_{X_1, \ldots, X_n | \sum_i X_i = s}(x_1, \ldots, x_n) = \frac{\mathbb{1}_{\{s\}}(\sum_i x_i)}{\binom{n}{s}}.$$

However, for $0 < s < n$, this constant mass function cannot be reconciled with a generative process (2.1) where a probability parameter θ is sampled from a probability measure (Q), followed by a sample of n independent Bernoulli(θ) trials $X_1, \ldots, X_n$; a degenerate value of $\theta \in \{0, 1\}$ would not admit $\sum X_i = s$, whilst any non-degenerate value $0 < \theta < 1$ would admit positive probability to $\sum X_i \neq s$.

Solution 2.2 *Predictive distribution for exchangeable binary sequences.* The result follows from substituting Theorem 2.1 into the conditional probability identity

$$\mathbb{P}_{X_{m+1}, \ldots, X_n | x_1, \ldots, x_m}(x_{m+1}, \ldots, x_n) = \frac{\mathbb{P}_{X_1, \ldots, X_n}(x_1, \ldots, x_n)}{\mathbb{P}_{X_1, \ldots, X_m}(x_1, \ldots, x_m)}.$$

Solution 2.3 *Variances under transformations.* If $\theta \sim \mathrm{Gamma}(a, b)$, then $\theta^{-1} \sim \mathrm{Inverse\text{-}Gamma}(a, b)$. The variances for these respective distributions are a/b^2 and $b^2/\{(a-1)^2(a-2)\}$ for $a > 2$. Consequently, either large a or small b increase the variance of θ, but reduce the variance of $1/\theta$.

Solution 2.4 *Asymptotic normality.* Let $\dot{x} = \sum_{i=1}^{n} x_i$ and $\bar{x} = \dot{x}/n$. Then

$$\sum_{i=1}^{n} \log F(x_i; \theta) = \dot{x} \log \theta + (n - \dot{x}) \log(1 - \theta)$$

$$\implies \frac{d}{d\theta} \log F(x_i; \theta) = \frac{\dot{x}}{\theta} - \frac{n - \dot{x}}{1 - \theta}$$

$$\implies I_n(\theta) = -\frac{d^2}{d\theta^2} \log F(x_i; \theta) = \frac{\dot{x}}{\theta^2} + \frac{n - \dot{x}}{(1 - \theta)^2}. \tag{B.1}$$

Setting the first derivative (B.1) equal to zero yields $\hat{\theta}_n = \dot{x}/n = \bar{x}$. Similarly for $Q(\theta) = \text{Beta}(\theta \mid a, b)$, the prior mode and $m_0 = (a-1)/(a+b-2)$ and

$$I_0(\theta) = -\frac{d^2}{d\theta^2} \log dQ(\theta) = \frac{a-1}{\theta^2} + \frac{b-1}{(1-\theta)^2}$$

$$\implies H_n = I_0(m_0) + I_n(\hat{\theta}_n)$$

$$\implies H_n^{-1} = \frac{(a-1)(b-1)\bar{x}(1-\bar{x})}{(a+b-2)^3\bar{x}(1-\bar{x}) + (a-1)(b-1)n}$$

$$\implies m_n = H_n^{-1}(I_0(m_0)m_0 + I_n(\hat{\theta}_n)\hat{\theta}_n) = \frac{(a-1)\bar{x}\{(a+b-2)^2(1-\bar{x}) + (b-1)n\}}{(a+b-2)^3\bar{x}(1-\bar{x}) + (a-1)(b-1)n}.$$

Asymptotically, as $n \to \infty$,

$$\theta \mid x_1, \ldots, x_n \overset{\cdot}{\sim} \text{Normal}\left(m_n, H_n^{-1}\right) \to \text{Normal}\left(\bar{x}, \frac{\bar{x}(1-\bar{x})}{n}\right).$$

Alternatively, from Section A.2, $Q(\theta \mid x_1, \ldots, x_n) = \text{Beta}(\theta \mid a + \dot{x}, b + n - \dot{x})$. The beta distribution is known to be approximately normal when both parameters grow large. Using the moments of the beta distribution, approximately

$$\theta \mid x_1, \ldots, x_n \overset{\cdot}{\sim} \text{Normal}\left(\frac{a+\dot{x}}{a+b+n}, \frac{(a+\dot{x})(b+n-\dot{x})}{(a+b+n)^2(a+b+n+1)}\right) \to \text{Normal}\left(\bar{x}, \frac{\bar{x}(1-\bar{x})}{n}\right).$$

Solution 3.1 *Identifying parents and children.*

$$\text{parents}(X_1) = \emptyset, \qquad\qquad \text{children}(X_1) = \{X_2, X_4\};$$
$$\text{parents}(X_2) = \{X_1\}, \qquad\qquad \text{children}(X_2) = \{X_4\};$$
$$\text{parents}(X_3) = \{X_4\}, \qquad\qquad \text{children}(X_3) = \emptyset;$$
$$\text{parents}(X_4) = \{X_1, X_2\}, \qquad\qquad \text{children}(X_4) = \{X_3\}.$$

Solution 3.2 *Identifying neighbours.* $\text{neighbours}(X_1) = \{X_2, X_4\}$, $\text{neighbours}(X_2) = \{X_1, X_4\}$, $\text{neighbours}(X_3) = \{X_4\}$, $\text{neighbours}(X_4) = \{X_1, X_2, X_3\}$.

Solution 3.3 *Identifying cliques.* $\{X_1, X_2, X_4\}$ and $\{X_3, X_4\}$ are both maximal cliques.

Solution 3.4 *Identifying separating sets.* $\{X_4\}$ separates $\{X_1, X_2\}$ from $\{X_3\}$; $\{X_1, X_4\}$ separates $\{X_2\}$ from $\{X_3\}$; $\{X_2, X_4\}$ separates $\{X_1\}$ from $\{X_3\}$.

Solution 3.5 *Belief network distribution.* $\mathbb{P}_\mathcal{G}(X_1, X_2, X_3, X_4) = \mathbb{P}(X_1)\,\mathbb{P}(X_2 \mid X_1)\,\mathbb{P}(X_4 \mid X_1, X_2)\,\mathbb{P}(X_3 \mid X_4)$.

Solution 3.6 *Identifying colliders.* The paths in (a) and (b) have no colliders, but in path (c) the node X_2 is a collider.

Solution 3.7 *Identifying d-separated and d-connected nodes.* X_1 and X_3 are *d*-separated by X_2 in (a) and (b) of Fig. 3.3, and *d*-connected by X_2 in (c).

Solution 3.8 *Identifying conditional independencies in a belief network.*

(i) $X_1 \not\perp\!\!\!\perp X_3$ in (a) and (b), and $X_1 \perp\!\!\!\perp X_3$ in (c).
(ii) $X_1 \perp\!\!\!\perp X_3 \mid X_2$ in (a) and (b), and $X_1 \not\perp\!\!\!\perp X_3 \mid X_2$ in (c).

Solution 3.9 *Markov network distribution.*

$$\mathbb{P}_G(X_1, X_2, X_3, X_4) = \phi_1(X_1, X_2, X_4)\phi_2(X_3, X_4).$$

Solution 3.10 *Pairwise Markov network distribution.*

$$\mathbb{P}_G(X_1, X_2, X_3, X_4) = \phi_{1,2}(X_1, X_2)\phi_{1,4}(X_1, X_4)\phi_{2,4}(X_2, X_4)\phi_{3,4}(X_3, X_4).$$

Solution 3.11 *Gaussian Markov random field.* Suppose $X = (X_1, \ldots, X_n) \sim N_n$ (μ, Σ), and let $\Lambda = \Sigma^{-1}$. Without loss of generality, to simplify notation assume $\mu = 0$.

For $x \in \mathbb{R}^n$, let $x_{-\ell} = (x_1, \ldots, x_{\ell-1}, x_{\ell+1}, \ldots, x_n)$ be the $(n-1)$-vector with component ℓ removed. From the density function of the multivariate normal distribution,

$$f(x_\ell \mid x_{-\ell}) \propto f(x_1, \ldots, x_n) \propto= e^{-\frac{1}{2}\sum_{i=1}^n \sum_{j=1}^n x_i \Lambda_{ij} x_j} \propto e^{-\frac{1}{2}\Lambda_{\ell\ell} x_\ell^2 - \sum_{j \neq \ell} \Lambda_{\ell j} x_\ell x_j}$$

when considered as a function of x_ℓ. The components x_j which affect this density are those j for which $\Lambda_{\ell j} \neq 0$, which by construction are those j for which $(\ell, j) \in E$. Hence

$$f(x_\ell \mid x_{-\ell}) = f(x_\ell \mid \text{neighbours}_G(x_\ell)).$$

Solution 4.1 *Conjugacy of Bernoulli and beta distributions.* Under the Bernoulli likelihood model,

$$p(\mathbf{x} \mid \theta) = \theta^{\dot{x}}(1 - \theta)^{n-\dot{x}},$$

where $\dot{x} = \sum_{i=1}^n x_i$. If $\theta \sim \text{Beta}(a, b)$ then

$$p(\theta) \propto \theta^{a-1}(1 - \theta)^{b-1}.$$

By (4.2),

$$\pi(\theta) \propto p(\mathbf{x} \mid \theta)\, p(\theta) \propto \theta^{a+\dot{x}-1}(1 - \theta)^{b+n-1},$$

which is proportional to the density of $\text{Beta}(a + \dot{x}, b + n - \dot{x})$.

Hence $\theta \mid \mathbf{x} \sim \text{Beta}(a + \dot{x}, b + n - \dot{x})$.

Solution 4.2 *Conjugacy of Poisson and gamma distributions.* Under the Poisson likelihood model,

$$p(\mathbf{x} \mid \theta) \propto \theta^{\dot{x}} e^{-n\theta},$$

where $\dot{x} = \sum_{i=1}^{n} x_i$. If $\theta \sim \text{Gamma}(a, b)$ then

$$p(\theta) \propto \theta^{a-1} e^{-b\theta}.$$

By (4.2),

$$\pi(\theta) \propto p(\mathbf{x} \mid \theta) \, p(\theta) \propto \theta^{a+\dot{x}-1} e^{-(b+n)\theta},$$

which is proportional to the density of $\text{Gamma}(a + \dot{x}, b + n)$.

Hence $\theta \mid \mathbf{x} \sim \text{Gamma}(a + \dot{x}, b + n)$.

Solution 4.3 *Conjugacy of uniform and Pareto distributions.* Under the uniform likelihood model, for $x_1, \ldots, x_n > 0$,

$$p(\mathbf{x} \mid \theta) = \frac{\prod_{i=1}^{n} \mathbb{1}_{[0,\theta]}(x_i)}{\theta^n} = \frac{\mathbb{1}_{[0,\theta]}(x_{(n)})}{\theta^n} = \frac{\mathbb{1}_{[x_{(n)},\infty)}(\theta)}{\theta^n},$$

where $x_{(n)} = \max\{x_1, \ldots, x_n\}$. If $\theta \sim \text{Pareto}(a, b)$ then

$$p(\theta) \propto \frac{\mathbb{1}_{[b,\infty)}(\theta)}{\theta^{a+1}}.$$

By (4.2),

$$\pi(\theta) \propto p(\mathbf{x} \mid \theta) \, p(\theta) \propto \frac{\mathbb{1}_{[b,\infty)}(\theta)\mathbb{1}_{[x_{(n)},\infty)}(\theta)}{\theta^{a+n+1}} = \frac{\mathbb{1}_{[\max\{b,x_{(n)}\},\infty)}(\theta)}{\theta^{a+n+1}},$$

which is proportional to the density of $\text{Pareto}(a + n, \max\{b, x_{(n)}\})$.

Hence $\theta \mid \mathbf{x} \sim \text{Pareto}(a + n, \max\{b, x_{(n)}\})$.

Solution 4.4 *Conjugacy of exponential and gamma distributions.* Under the exponential likelihood model,

$$p(\mathbf{x} \mid \theta) = \theta^n e^{-\theta \dot{x}},$$

where $\dot{x} = \sum_{i=1}^{n} x_i$. If $\theta \sim \text{Gamma}(a, b)$ then

$$p(\theta) \propto \theta^{a-1} e^{-b\theta}.$$

By (4.2),

$$\pi(\theta) \propto p(\mathbf{x} \mid \theta) \, p(\theta) \propto \theta^{a+n-1} e^{-(b+\dot{x})\theta},$$

which is proportional to the density of $\text{Gamma}(a + n, b + \dot{x})$.

Hence $\theta \mid \mathbf{x} \sim \text{Gamma}(a + n, b + \dot{x})$.

Solution 4.5 *Calculating a marginal distribution.* For $\theta_1, \theta_2 > 0$,

$$\pi(\theta_1) = \frac{b^a \theta_1^a e^{-b\theta_1}}{\Gamma(a)} \int_0^\infty e^{-\theta_1 \theta_2} \, d\theta_2 = \frac{b^a \theta_1^a e^{-b\theta_1}}{\Gamma(a)} \frac{-e^{-\theta_1 \theta_2}}{\theta_1} \Big|_0^\infty$$

$$= \frac{b^a \theta_1^{a-1} e^{-b\theta_1}}{\Gamma(a)}$$

and

$$\pi(\theta_2) = \frac{b^a}{\Gamma(a)} \int_0^\infty \theta_1^a e^{-(b+\theta_2)\theta_1} \, d\theta_1 = \frac{b^a}{\Gamma(a)(b+\theta_2)^{a+1}} \int_0^\infty x^a e^{-x} \, dx$$

$$= \frac{\Gamma(a+1)b^a}{\Gamma(a)(b+\theta_2)^{a+1}} = \frac{a \, b^a}{(b+\theta_2)^{a+1}}.$$

Solution 4.6 *Credible interval for the exponential distribution.*

$$\int_{-\infty}^{\theta_*} \pi(\theta) \, d\theta = \int_0^{\theta_*} \lambda e^{-\lambda\theta} \, d\theta = -e^{-\lambda\theta} \Big|_0^{\theta_*} = 1 - e^{-\lambda\theta_*}.$$

Then $1 - e^{-\lambda\theta_*} = (1-\alpha)/2 \iff \theta_* = -\log\{(1+\alpha)/2\}/\lambda$. Similarly,

$$\int_{\theta^*}^\infty \pi(\theta) \, d\theta = \int_{\theta^*}^\infty \lambda e^{-\lambda\theta} \, d\theta = -e^{-\lambda\theta} \Big|_{\theta^*}^\infty = e^{-\lambda\theta^*}$$

and $e^{-\lambda\theta^*} = (1-\alpha)/2 \iff \theta^* = -\log\{(1-\alpha)/2\}/\lambda$.
Hence a $100\alpha\%$ credible interval for θ is

$$[-\log\{(1+\alpha)/2\}/\lambda, \, -\log\{(1-\alpha)/2\}/\lambda].$$

Solution 5.1 *Monte Carlo probabilities.* The probability of θ lying inside $A \subset \Theta$ can be expressed as an expectation using the indicator function,

$$\mathbb{P}_\pi(\theta \in A) = \int_A \pi(\theta) \, d\theta = \int_\Theta \pi(\theta) \, \mathbb{1}_A(\theta) \, d\theta = \mathbb{E}_\pi\{\mathbb{1}_A(\theta)\}.$$

Hence a Monte Carlo estimate for $\mathbb{P}_\pi(\theta \in A)$ can be obtained by

$$\hat{\mathbb{P}}_\pi(\theta \in A) = \frac{1}{M} \sum_{i=1}^M \mathbb{1}_A(\theta^{(i)}).$$

Solution 5.2 *Monte Carlo estimate of a conditional expectation.* Using the identity,

$$\mathbb{E}_{\pi|A}(g(\theta) \mid \theta \in A) = \mathbb{E}_\pi\{\mathbb{1}_A(\theta) \, g(\theta)\}/ \mathbb{E}_\pi\{\mathbb{1}_A(\theta)\},$$

it follows that conditional expectations can be approximated by

$$\mathbb{E}_{\pi|A}(g(\theta) \mid \theta \in A) = \frac{\sum_{i=1}^{M} \mathbb{1}_A(\theta^{(i)}) g(\theta^{(i)})}{\sum_{i=1}^{M} \mathbb{1}_A(\theta^{(i)})}$$

provided $\sum_{i=1}^{M} \mathbb{1}_A(\theta^{(i)}) > 0$ (meaning there are samples lying in A).

Solution 5.3 *Monte Carlo credible interval*. From Exercise 5.1, it follows that

$$\hat{\mathbb{P}}_\pi(\theta \in [\theta_{(M(1-\alpha)/2)}, \theta_{(M(1+\alpha)/2)}]) = \alpha$$

and therefore $R_\alpha = [\theta_{(M(1-\alpha)/2)}, \theta_{(M(1+\alpha)/2)}]$ is a Monte Carlo approximated $100\alpha\%$ credible region for θ.

Solution 5.4 *Monte Carlo optimal decision estimation*. The Monte Carlo estimate of the expected loss function is

$$\hat{\mathbb{E}}_\pi\{\ell(\hat{\theta},\theta)\} = -\frac{1}{3} \sum_{i=1}^{3} \exp\{-(\hat{\theta} - \theta^{(i)})^2/10\},$$

which is plotted below.

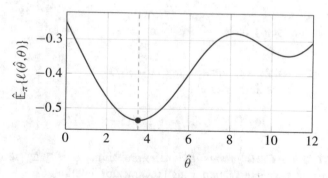

The following Python code then identifies the minimising value numerically using the *SciPy*[1] library function `scipy.optimize.minimize.`.

```
#! /usr/bin/env python
## gaussian_loss_example.py
import numpy as np
from scipy.optimize import minimize

def E_loss(y,z):
    return(sum([-np.exp(-.1*(y-v)**2) for v in z]))

z = [2,5,11]
print("Minimiser =",minimize(E_loss,np.mean(z),z).x[0])
```

```
Minimiser = 3.5321781359647586
```

---

[1] https://www.scipy.org.

Hence, an approximate Bayesian estimate of θ under this loss function is $\hat{\theta} \approx 3.532$. [This estimate differs from the Monte Carlo estimate of the mean of π, $\hat{\mathbb{E}}_{\pi}(\theta) = (2 + 5 + 11)/3 = 6$.]

Solution 5.5 *Importance sampling Monte Carlo standard error.* Using the identity (5.8), it follows from (5.4) that

$$\text{s.e.}\{\hat{\mathbb{E}}_{\pi}^{\text{IS}}\{g(\theta)\}\} = \sqrt{\frac{1}{M(M-1)} \sum_{i=1}^{M} \left\{ w_i g(\theta^{(i)}) - \frac{1}{m} \sum_{i=1}^{M} w_i g(\theta^{(i)}) \right\}^2}.$$

Solution 5.6 *Gibbs sampling.*

(i) $\pi(\theta_1, \theta_2) = \frac{1}{2}\phi(\theta_1 - \mu)\phi(\theta_2 - \mu) + \frac{1}{2}\phi(\theta_1 + \mu)\phi(\theta_2 + \mu)$.

(ii)

$$\pi(\theta_2) = \int \pi(\theta_1, \theta_2) \, d\theta_1 = \frac{1}{2}\phi(\theta_2 - \mu) + \frac{1}{2}\phi(\theta_2 + \mu)$$

$$\implies \pi(\theta_1 \mid \theta_2) = \frac{\pi(\theta_1, \theta_2)}{\pi(\theta_2)} = \frac{\phi(\theta_1 - \mu)\phi(\theta_2 - \mu) + \phi(\theta_1 + \mu)\phi(\theta_2 + \mu)}{\phi(\theta_2 - \mu) + \phi(\theta_2 + \mu)}$$

$$= w(\theta_2) \, \phi(\theta_1 - \mu) + \{1 - w(\theta_2)\} \, \phi(\theta_1 + \mu),$$

where

$$w(\theta_i) = \frac{\phi(\theta_i - \mu)}{\phi(\theta_i - \mu) + \phi(\theta_i + \mu)}$$

$$= (1 + e^{-2\theta_i \mu})^{-1}.$$

By symmetry,

$$\pi(\theta_2 \mid \theta_1) = w(\theta_1) \, \phi(\theta_2 - \mu) + \{1 - w(\theta_1)\} \, \phi(\theta_2 + \mu).$$

(iii) As μ increases, the target density becomes bimodal and the mixture weight $w(\theta_i) \to 0$ if θ_i is negative, and $w(\theta_i) \to 1$ if θ_i is positive, and therefore θ becomes stuck near either $(-\mu, -\mu)$ or (μ, μ).

Solution 5.7 *Gibbs sampling implementation.*

```
#! /usr/bin/env python
## gibbs_sampling_2d.py
import numpy as np
import matplotlib.pyplot as plt

def full_conditional(y,mu,g):
    z = g.binomial(1,1/(1+np.exp(-2*y*mu)))
    return(g.normal() + (mu if z else -mu))

def gibbs_sampling(M=100,seed=9,initial=[0,0],mu=1):
    gen = np.random.default_rng(seed=seed)
    xs = np.empty(shape=[M+1,2])
    xs[0,] = x = list(initial)
    for i in range(M):
        for j in range(2):
            x[j] = full_conditional(x[1-j],mu,gen)
        xs[i+1,] = x
    return(xs)

def trace_plots(z):
    fig,axs=plt.subplots(1,len(z),figsize=(12,4), constrained_layout=True)
    for ind in range(len(z)):
        x,y = z[ind][:,0],z[ind][:,1]
        axs[ind].plot(x,y,'bx-',linewidth=.2,markersize=4)
        axs[ind].set_xlabel(r'$\theta_1$', fontsize=16)
        axs[ind].set_ylabel(r'$\theta_2$', fontsize=16)
    plt.show()

trace_plots([gibbs_sampling(mu=mu) for mu in (1,3)])
```

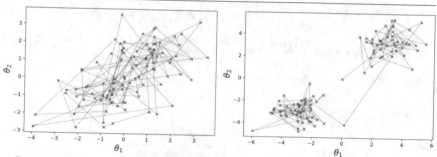

Starting the chain from $(0, 0)$, the left-hand plot shows good mixing when $\mu = 1$, whereas in right-hand plot when $\mu = 3$ there is only one transition between the two modes during 100 iterations.

Solution 5.8 *Detailed balance of Metropolis-Hastings algorithm.* If $\theta = \theta'$, then by symmetry (5.12) trivially holds. If $\theta \neq \theta'$, then (5.16) simplifies to $p(\theta' \mid \theta) = \alpha(\theta, \theta')q(\theta' \mid \theta)$ and it remains to show

$$\pi(\theta)\alpha(\theta, \theta')q(\theta' \mid \theta) = \pi(\theta')\alpha(\theta', \theta)q(\theta \mid \theta').$$

If $\pi(\theta') = 0$ then from (5.15), $\alpha(\theta, \theta') = 0$ and the equality holds. So now suppose $\pi(\theta') > 0$,

$$\pi(\theta)q(\theta' \mid \theta)\alpha(\theta, \theta') = \pi(\theta)q(\theta' \mid \theta) \min \left\{ 1, \frac{\pi(\theta')q(\theta \mid \theta')}{\pi(\theta)q(\theta' \mid \theta)} \right\}$$

$$= \min\{\pi(\theta)q(\theta' \mid \theta), \pi(\theta')q(\theta \mid \theta')\}$$

$$= \pi(\theta')q(\theta \mid \theta') \min \left\{ 1, \frac{\pi(\theta)q(\theta' \mid \theta)}{\pi(\theta')q(\theta \mid \theta')} \right\}$$

$$= \pi(\theta')q(\theta \mid \theta')\alpha(\theta', \theta).$$

Solution 5.9 *Gibbs sampling as Metropolis-Hastings special case.* Then the ratio of posterior densities when $\theta'_{-j} = \theta_{-j}$ is

$$\frac{\pi(\theta')}{\pi(\theta)} = \frac{\pi(\theta_j \mid \theta_{-j})\pi(\theta_{-j})}{\pi(\theta'_j \mid \theta_{-j})\pi(\theta_{-j})} = \frac{\pi(\theta_j \mid \theta_{-j})}{\pi(\theta'_j \mid \theta_{-j})},$$

which cancels with the ratio of proposal densities in the Metropolis-Hastings acceptance probability (5.15), and hence $\alpha(\theta, \theta') = 1$ and all such Metropolis-Hastings proposals are accepted with probability 1.

Solution 5.10 *Metropolis-Hastings implementation.*

```python
#! /usr/bin/env python
## mh_rw_sampling_2d.py
import numpy as np
import matplotlib.pyplot as plt

def bi_gauss(y,mu):
    return(np.exp(-.5*np.dot(y-mu,y-mu)))

def target(y,mu):
    return(np.log(bi_gauss(y,mu)+bi_gauss(y,-mu)))

def mh_sampling(M=100,seed=9,initial=[0,0],mu=1):
    gen = np.random.default_rng(seed=seed)
    xs = np.empty(shape=[M+1,2])
    xs[0,] = x = np.array(initial)
    p = target(x,mu)
    acc = 0
    for i in range(M):
        x_ = x + 2*gen.normal(size=2)
        p_ = target(x_,mu)
        if p_ > p or -gen.exponential() < p_ - p:
            x,p = x_,p_
            acc +=1
        xs[i+1,] = x
    print("Mu="+str(mu),"acc. rate:",str(acc/M*100)+"%")
    return(xs)

def trace_plots(z):
    fig,axs=plt.subplots(1,len(z),figsize=(12,4), constrained_layout=True)
    for ind in range(len(z)):
        x,y = z[ind][:,0],z[ind][:,1]
        axs[ind].plot(x,y,'bx-',linewidth=.2,markersize=4)
        axs[ind].set_xlabel(r'$\theta_1$', fontsize=16)
        axs[ind].set_ylabel(r'$\theta_2$', fontsize=16)
    plt.show()

trace_plots([mh_sampling(mu=mu) for mu in (1,3)])
```

```
Mu=1 acc. rate: 32.0%
Mu=3 acc. rate: 33.0%
```

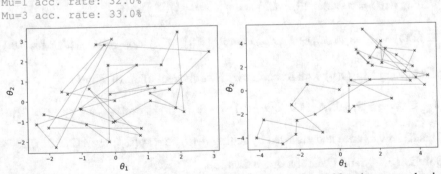

In comparison with Gibbs sampling, there are fewer than 100 unique samples in each case.

Solution 5.11 *ELBO equivalence.* Since $\log p(\mathbf{x}, \theta) = \log \pi(\theta) + \log p(\mathbf{x})$,

$$
\begin{aligned}
\mathrm{KL}(q(\theta) \parallel \pi(\theta)) &= \int_{\Theta} q(\theta) \log \frac{q(\theta)}{\pi(\theta)} \, \mathrm{d}\theta \\
&= \int_{\Theta} q(\theta) \log q(\theta) \, \mathrm{d}\theta - \int_{\Theta} q(\theta) \log p(\mathbf{x}, \theta) \, \mathrm{d}\theta + \int_{\Theta} q(\theta) \log p(\mathbf{x}) \, \mathrm{d}\theta \\
&= \mathbb{E}_q \log q(\theta) - \mathbb{E}_q \log p(\mathbf{x}, \theta) + \log p(\mathbf{x}) \\
&= -\mathrm{ELBO}(q) + \log p(\mathbf{x}).
\end{aligned}
$$

The $\log p(\mathbf{x})$ term does not depend on the density q, and so minimising this expression corresponds to maximising $\mathrm{ELBO}(q)$.

Solution 5.12 *ELBO identity.* Since $\log p(\mathbf{x}, \theta) = \log p(\mathbf{x} \mid \theta) + \log p(\theta)$,

$$
\begin{aligned}
\mathrm{ELBO}(q) &= \mathbb{E}_q \log p(\mathbf{x}, \theta) - \mathbb{E}_q \log q(\theta) \\
&= \mathbb{E}_q \log p(\mathbf{x} \mid \theta) + \mathbb{E}_q \log p(\theta) - \mathbb{E}_q \log q(\theta) \\
&= \mathbb{E}_q \log p(\mathbf{x} \mid \theta) - \mathbb{E}_q \log \frac{q(\theta)}{p(\theta)} \\
&= \mathbb{E}_q \log p(\mathbf{x} \mid \theta) - \mathrm{KL}(q(\theta) \parallel p(\theta)).
\end{aligned}
$$

Solution 5.13 *CAVI derivation.* Using the identity $p(\mathbf{x}, \theta) = \pi(\theta) p(\mathbf{x})$,

$$
\begin{aligned}
\mathrm{ELBO}(q) &= \mathbb{E}_q \log p(\mathbf{x}, \theta) - \mathbb{E}_q \log q(\theta) \\
&= \mathbb{E}_q \log \pi(\theta) + \mathbb{E}_q \log p(\mathbf{x}) - \mathbb{E}_q \log q(\theta).
\end{aligned}
$$

Since $p(\mathbf{x})$ does not depend on q, maximising $\mathrm{ELBO}(q)$ is equivalent to maximising

$$
\widetilde{\mathrm{ELBO}}(q) = \mathbb{E}_q \log \pi(\theta) - \mathbb{E}_q \log q(\theta).
$$

Writing $\pi(\theta) = \pi(\theta_{-j}) \pi(\theta_j \mid \theta_{-j})$ and $q(\theta) = q_{-j}(\theta_{-j}) q_j(\theta_j)$,

$$
\widetilde{\mathrm{ELBO}}(q) = \mathbb{E}_{q_{-j}} \log \pi(\theta_{-j}) + \mathbb{E}_{q_j} \mathbb{E}_{q_{-j}} \log \pi(\theta_j \mid \theta_{-j}) - \mathbb{E}_{q_{-j}} \log q_{-j}(\theta_{-j}) - \mathbb{E}_{q_j} \log q_j(\theta_j).
$$

Maximising $\widetilde{\mathrm{ELBO}}(q)$ with respect to q_j is equivalent to maximising

$$
\mathbb{E}_{q_j} \mathbb{E}_{q_{-j}} \log \pi(\theta_j \mid \theta_{-j}) - \mathbb{E}_{q_j} \log q_j(\theta_j) = -\mathrm{KL}[q_j(\theta_j) \parallel \exp\{\mathbb{E}_{q_{-j}} \log \pi(\theta_j \mid \theta_{-j})\}].
$$

This KL-divergence is minimised by setting $q_j(\theta_j) \propto \exp\{\mathbb{E}_{q_{-j}} \log \pi(\theta_j \mid \theta_{-j})\}$.

Solution 5.14 *CAVI Gaussian approximation.*

(i) With just two components, $\theta_{-j} = \theta_{\bar{j}}$. Taking the conditional distribution of a bivariate normal,

$$\pi(\theta_j \mid \theta_{\bar{\jmath}}) = \text{Normal}\left(\mu_j + \frac{\Sigma_{j\bar{\jmath}}}{\Sigma_{\bar{\jmath}\bar{\jmath}}}(\theta_{\bar{\jmath}} - \mu_{\bar{\jmath}}), \ \Sigma_{jj} - \frac{\Sigma_{j\bar{\jmath}}^2}{\Sigma_{\bar{\jmath}\bar{\jmath}}}\right)$$

$$\implies \pi(\theta_j \mid \theta_{\bar{\jmath}}) \propto \exp\left\{-\frac{1}{2s_j^2}\left(\theta_j - \mu_j - \frac{\Sigma_{j\bar{\jmath}}}{\Sigma_{\bar{\jmath}\bar{\jmath}}}(\theta_{\bar{\jmath}} - \mu_{\bar{\jmath}})\right)^2\right\}$$

$$\implies \log \pi(\theta_j \mid \theta_{\bar{\jmath}}) = -\frac{1}{2s_j^2}\left(\theta_j - \mu_j - \frac{\Sigma_{j\bar{\jmath}}}{\Sigma_{\bar{\jmath}\bar{\jmath}}}(\theta_{\bar{\jmath}} - \mu_{\bar{\jmath}})\right)^2 + \text{constant}$$

$$\implies \mathbb{E}_{q_j} \log \pi(\theta_j \mid \theta_{\bar{\jmath}}) = -\frac{1}{2s_j^2}\left\{\theta_j^2 - 2\theta_j\left(\mu_j - \frac{\Sigma_{j\bar{\jmath}}}{\Sigma_{\bar{\jmath}\bar{\jmath}}}(m_{\bar{\jmath}} - \mu_{\bar{\jmath}})\right)\right\} + \text{constant}$$

$$= -\frac{1}{2s_j^2}(\theta_j^2 - 2\theta_j m_j) + \text{constant}$$

$$\implies \exp\{\mathbb{E}_{q_j} \log \pi(\theta_j \mid \theta_{\bar{\jmath}})\} \propto \exp\left\{-\frac{1}{2s_j^2}\left(\theta_j^2 - 2\theta_j m_j\right)\right\} \propto \exp\left\{-\frac{1}{2s_j^2}\left(\theta_j - m_j\right)^2\right\}.$$

(ii) The variances of the component densities q_j are fixed, and the algorithm will converge when each mean value $m_j = \mu_j$.

(iii) The following Python code implements coordinate ascent variational inference for the bivariate normal distribution with correlation .95. The printed output gives the fitted mean and variance values. The contour plot mirrors Fig. 5.2(a).

```
#! /usr/bin/env python
## cavi.py
import numpy as np
from scipy.stats import multivariate_normal
import matplotlib.pyplot as plt

#target distribution parameters
mu, Sigma = [0,0], [[1,.95], [.95,1]]

#initial approximation
m, s = [2,3], [.1,9]

for _ in range(200):
    for j in range(2):
        m[j]=mu[j]+Sigma[j][1-j]/Sigma[1-j][1-j]*(m[1-j]-mu[1-j])
        s[j]=Sigma[j][j]-Sigma[j][1-j]**2/Sigma[1-j][1-j]

print(m,s)

x,y = np.mgrid[-2.5:2.5:.05, -2.5:2.5:.05]
xy = np.dstack((x,y))
N1 = multivariate_normal(mu,Sigma)
N2 = multivariate_normal(m,[[s[0],0],[0,s[1]]])
plt.rcParams["figure.figsize"] = (5,5)
plt.contour(x,y,N1.pdf(xy),10,colors='black')
plt.contour(x,y,N2.pdf(xy),10,colors='blue')
plt.xlabel(r'$\theta_1$', fontsize=14)
plt.ylabel(r'$\theta_2$', fontsize=14)
plt.gcf().canvas.manager.set_window_title('Density contours')
plt.show()
```

```
[3.880071824689807e-09, 3.6860682334553167e-09] [0.09750000000000003,
0.09750000000000003]
```

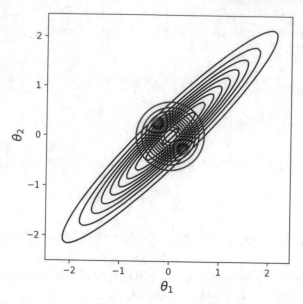

Solution 7.1 *Bayes factors for Gaussian distributions.*

(i) Let $\ddot{x} = \sum_{i=1}^{n} x_i^2$, $\dot{x} = \sum_{i=1}^{n} x_i$. From Section A.3, under model M_1,

$$p_1(\mathbf{x}) = \frac{\exp[-\frac{1}{2}\{\ddot{x} - (\sigma^{-2} + n)^{-1}\dot{x}^2\}]}{(2\pi)^{\frac{n}{2}}(n\sigma^2 + 1)^{\frac{1}{2}}}.$$

Similarly,

$$p_1(\mathbf{y}) = \frac{\exp[-\frac{1}{2}\{\ddot{y} - (\sigma^{-2} + n)^{-1}\dot{y}^2\}]}{(2\pi)^{\frac{n}{2}}(n\sigma^2 + 1)^{\frac{1}{2}}}.$$

Under model M_0

$$p_0(\mathbf{x}, \mathbf{y}) = \frac{\exp[-\frac{1}{2}\{\ddot{x} + \ddot{y} - (\sigma^{-2} + 2n)^{-1}(\dot{x} + \dot{y})^2\}]}{(2\pi)^{n}(2n\sigma^2 + 1)^{\frac{1}{2}}}.$$

Consequently, the Bayes factor in favour of M_0 is

$$B_{01}(\mathbf{x}, \mathbf{y}) = \frac{p_0(\mathbf{x}, \mathbf{y})}{p_1(\mathbf{x}) p_1(\mathbf{y})}$$

$$= \frac{(n\sigma^2 + 1)}{(2n\sigma^2 + 1)^{\frac{1}{2}}} \exp\{-\frac{1}{2}(\sigma^{-2} + n)^{-1}(\dot{x}^2 + \dot{y}^2) + \frac{1}{2}(\sigma^{-2} + 2n)^{-1}(\dot{x} + \dot{y})^2\}.$$

(ii) From the previous expression,

$$B_{01}(\mathbf{x}, \mathbf{y}) = \sqrt{1 + \frac{n^2\sigma^4}{2n\sigma^2 + 1}} \exp\{-\frac{1}{2}(\sigma^{-2} + n)^{-1}(\dot{x}^2 + \dot{y}^2) + \frac{1}{2}(\sigma^{-2} + 2n)^{-1}(\dot{x} + \dot{y})^2\}.$$

For large σ,

$$B_{01}(\mathbf{x}, \mathbf{y}) \approx \sigma\left(\tfrac{n}{2}\right)^{\frac{1}{2}} \exp\{-\tfrac{1}{2n}(\dot{x}^2 + \dot{y}^2) + \tfrac{1}{4n}(\dot{x} + \dot{y})^2\}$$
$$= \sigma\left(\tfrac{n}{2}\right)^{\frac{1}{2}} \exp\{-\tfrac{1}{4n}(\dot{x} - \dot{y})^2\}.$$

Clearly $B_{01}(\mathbf{x}, \mathbf{y}) \to \infty$ as $\sigma \to \infty$. This means the simpler model, where $\theta_X = \theta_Y$, will *always* be preferred.

Solution 7.2 *BIC for Gaussian distributions.*

(i) It is easily shown that the maximum likelihood estimates for the mean parameters θ_X and θ_Y under the two models are the corresponding sample means:

$$M_0 : \hat{\theta}_X = \hat{\theta}_Y = \frac{\bar{x} + \bar{y}}{2};$$
$$M_1 : \hat{\theta}_X = \bar{x}, \quad \hat{\theta}_Y = \bar{y},$$

where $\bar{x} = \frac{1}{n}\sum_{i=1}^{n} x_i$, $\bar{y} = \frac{1}{n}\sum_{i=1}^{n} y_i$. Then for model M_1,

$$\log p_1(\mathbf{x}, \mathbf{y} \mid \hat{\theta}_X, \hat{\theta}_Y) = \log p_1(\mathbf{x} \mid \hat{\theta}_X) + \log p_1(\mathbf{y} \mid \hat{\theta}_Y)$$
$$= -n\log(2\pi) - \frac{1}{2}\sum_{i=1}^{n}(x_i - \bar{x})^2 - \frac{1}{2}\sum_{i=1}^{n}(y_i - \bar{y})^2$$
$$\implies \text{BIC}_1 = 2n\log(2\pi) + \sum_{i=1}^{n}(x_i - \bar{x})^2 + \sum_{i=1}^{n}(y_i - \bar{y})^2 + 2\log(2n).$$

Under model M_0,

$$\log p_0(\mathbf{x}, \mathbf{y} \mid \hat{\theta}_X = \hat{\theta}_Y) = -n\log(2\pi) - \frac{1}{2}\sum_{i=1}^{n}\left\{\left(x_i - \frac{\bar{x} + \bar{y}}{2}\right)^2 + \left(y_i - \frac{\bar{x} + \bar{y}}{2}\right)^2\right\}$$
$$\implies \text{BIC}_0 = 2n\log(2\pi) + \sum_{i=1}^{n}\left\{\left(x_i + \frac{\bar{x} + \bar{y}}{2}\right)^2 + \left(y_i - \frac{\bar{x} + \bar{y}}{2}\right)^2\right\} + \log(2n).$$

(ii) The Bayes factor can be approximated using (7.1).

$$\text{BIC}_0 - \text{BIC}_1 = -n\bar{x}^2 - n\bar{y}^2 + \frac{n}{2}(\bar{x} + \bar{y})^2 - \log(2n)$$
$$= -\frac{n}{2}\bar{x}^2 - \frac{n}{2}\bar{y}^2 + n\bar{x}\bar{y} - \log(2n)$$
$$= -\frac{n}{2}(\bar{x} - \bar{y})^2 - \log(2n)$$
$$\implies B_{01} \approx \exp\left\{-\frac{1}{2}(\text{BIC}_0 - \text{BIC}_1)\right\} = \sqrt{2n}\,\exp\left\{\frac{n}{4}(\bar{x} - \bar{y})^2\right\}.$$

Solution 8.1 *Marginal density for regression coefficients.*

(i) Starting from the joint prior probability density function,

$$p(\beta, \sigma^2) = \frac{b^a \exp\left\{-\sigma^{-2}\left(b + \beta^{\mathsf{T}} V^{-1}\beta/2\right)\right\}}{(2\pi)^{\frac{p}{2}} |V|^{\frac{1}{2}} \Gamma(a) \, \sigma^{2(a-1)+p}}$$

$$\implies p(\beta) = \int_0^\infty p(\beta, \sigma^2) \, d\sigma^{-2} = \frac{b^a \, \Gamma(a + \frac{p}{2})}{(2\pi)^{\frac{p}{2}} |V|^{\frac{1}{2}} \Gamma(a) \, (b + \beta^{\mathsf{T}} V^{-1}\beta/2)^{a+\frac{p}{2}}},$$

by comparison of the integrand as a function of σ^{-2} with the probability density function for a Gamma$(a + p/2, b + \beta^{\mathsf{T}} V^{-1}\beta/2)$ random variable.

(ii) If $V = I_p$, then

$$p(\beta) = \frac{\Gamma(a + \frac{p}{2})}{(2\pi b)^{\frac{p}{2}} \Gamma(a)} \left(1 + \frac{\beta^{\mathsf{T}}\beta}{2b}\right)^{-(a+\frac{p}{2})}.$$

Regarded as a function of $\|\beta\|$, this density takes the same form as the density function of $t_{2a}(0, \frac{2b}{2a+p-1} I_p)$, except that $\|\beta\|$ can only take non-negative values.

Solution 8.2 *Linear model matrix inverse.* Setting $A = I_n$ and $U = W = X$, the matrix inversion lemma gives $(XVX^{\mathsf{T}} + I_n)^{-1} = I_n - X(V^{-1} + X^{\mathsf{T}}X)^{-1}X^{\mathsf{T}} = I_n - XV_nX^{\mathsf{T}}$.

Solution 8.3 *Linear model matrix determinant.* Setting $A = I_n$ and $U = W = X$, the matrix determinant lemma gives $|I_n + XVX^{\mathsf{T}}| = |V^{-1} + X^{\mathsf{T}}X||V|$.

Solution 8.4 *Linear model code.* The function `lm_log_likelihood` is one possible Python implementation:

```
## linear_regression.py

import numpy as np
from scipy.special import gammaln

def lm_log_likelihood(y,X,a=0.1,b=0.1,lam=0.01):
    n, p= X.shape
    XtX = np.dot(np.transpose(X),X)
    Xty = np.dot(np.transpose(X),y)
    V_n = np.linalg.inv(XtX+lam*np.identity(p))
    det_Vn= np.linalg.slogdet(V_n)[1]
    m_n = np.dot(V_n,Xty)
    a_n = a+.5*n
    b_n = b + .5*(np.dot(y,y) - np.dot(np.transpose(Xty),m_n))
    ml = .5*det_Vn + .5*p*np.log(lam) - a_n*np.log(b_n)
    + a*np.log(b)+gammaln(a_n)+.5*n*np.log(2*np.pi)-gammaln(a)
    return(ml,m_n)
```

Solution 8.5 *Orthogonal covariate matrix marginal likelihood.* If $V = \lambda^{-1} I_p$ and $X^{\mathsf{T}}X = I_p$, then

$$p(y \mid X) = \left(\frac{\lambda}{1+\lambda}\right)^{\frac{p}{2}} \frac{\Gamma(a+n/2) \, b^a}{(2\pi)^{\frac{n}{2}} \Gamma(a) \, (b + \frac{1}{2}y^{\mathsf{T}}y - \frac{1}{2(1+\lambda)}y^{\mathsf{T}}XX^{\mathsf{T}}y)^{a+\frac{n}{2}}}.$$

This expression does not require a matrix inversion, unlike the evaluation of the matrix V_n (8.10) required for the general case.

Solution 8.6 *Zellner's g-prior.* If $V = g(X^\mathsf{T}X)^{-1}$, then

$$p(y \mid X) = \frac{\Gamma(a + n/2)\, b^a}{(2\pi)^{\frac{n}{2}}\, \Gamma(a)\, (1 + g)^{\frac{p}{2}}\, (b + \frac{1}{2}y^\mathsf{T}y - \frac{g}{2(1+g)} y^\mathsf{T} X (X^\mathsf{T}X)^{-1} X^\mathsf{T} y)^{a + \frac{n}{2}}}.$$

Solution 9.1 *Dirichlet process marginal likelihood.* Let $\mathbf{x}' = (x_1', \ldots, x_k')$ be the $k \le n$ unique values which occur in $\mathbf{x}$, and let $n_j = \sum_{i=1}^n \mathbb{1}_{\{x_j'\}}(x_i)$ be the number of occurrences of x_j'. Also for $j = 1, \ldots, k$ let $B_j = \{x_j'\}$ and $B_{k+1} = \mathcal{X}/\cup_j B_j$. Then

$$p(\mathbf{x} \mid \mathbb{P}) = \prod_{j=1}^k \mathbb{p}(x_j')^{n_j} = \prod_{j=1}^k \mathbb{P}(B_j)^{n_j}$$

and

$$p(\mathbb{P}(B_1), \ldots, \mathbb{P}(B_{k+1})) = \frac{\Gamma(\alpha)}{\prod_{j=1}^k \Gamma\{\alpha\, \mathbb{P}_0(B_j)\}} \prod_{j=1}^k \mathbb{P}(B_j)^{\alpha\, \mathbb{P}_0(B_j)}$$

$$= \frac{\Gamma(\alpha)}{\prod_{j=1}^k \Gamma\{\alpha\, \mathbb{P}_0(x_j')\}} \prod_{j=1}^k \mathbb{P}(B_j)^{\alpha\, \mathbb{P}_0(x_j')}$$

$$\implies p(\mathbf{x}, \mathbb{P}(B_1), \ldots, \mathbb{P}(B_{k+1})) = \frac{\Gamma(\alpha)}{\prod_{j=1}^k \Gamma\{\alpha\, \mathbb{p}_0(x_j')\}} \prod_{j=1}^k \mathbb{P}(B_j)^{\alpha\, \mathbb{P}_0(x_j') + n_j}.$$

$$(B.2)$$

Marginalising (B.2) with respect to $(\mathbb{P}(B_1), \ldots, \mathbb{P}(B_{k+1}))$,

$$\implies p(\mathbf{x}) = \int p(\mathbf{x}, \mathbb{P}(B_1), \ldots, \mathbb{P}(B_{k+1}))\, d(\mathbb{P}(B_1), \ldots, \mathbb{P}(B_{k+1}))$$

$$= \frac{\Gamma(\alpha)}{\prod_{j=1}^k \Gamma\{\alpha\, \mathbb{p}_0(x_j')\}} \frac{\prod_{j=1}^k \Gamma\{\alpha\, \mathbb{p}_0(x_j') + n_j\}}{\Gamma(\alpha + n)},$$

by comparison with the normalising constant of the corresponding Dirichlet distribution.

Solution 9.2 *Dirichlet process sampling.* The following function `dirichlet_process_sample` is one possible Python implementation.

```
#! /usr/bin/env python
# dirichlet_process_sample.py

import numpy as np
import matplotlib.pyplot as plt
from scipy.stats import geom

def dirichlet_process_sample(a,P_0=geom,th=.1,m=50,seed=0):
    gen = np.random.default_rng(seed=seed)
    p = [P_0.pmf(k,th) for k in range(1,m)]+[P_0.sf(m+1,th)]
    return(range(1,m+1),gen.dirichlet(a*np.array(p)))
plt.plot(*dirichlet_process_sample(10),ls='None',marker='x')
plt.plot(*dirichlet_process_sample(1000),ls='None',marker='x')
plt.plot(*dirichlet_process_sample(100000),ls='None',marker='x')
plt.gcf().canvas.manager.set_window_title('Dirichlet process samples')
plt.show()
```

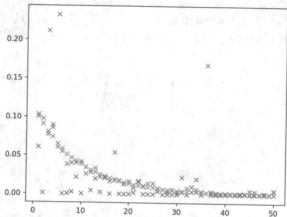

For larger values of α, the sampled mass functions more closely resemble the base geometric distribution.

Solution 9.3 *Binary partition index* For $x \in \mathbb{R}$, the index of x at any level m is obtained by calculating the m-digit binary representation of $F_0(x)$. This is achieved by the following Python code:

```
#! /usr/bin/env python
# binary_partition_index.py

from scipy.stats import norm

def binary_parition_index(F_0,x,m):
    b,y = [],F_0(x)
    for _ in range(m):
        b += [int(2*y)]
        y = 2*y - b[-1]
    print(''.join(map(str,b)), end='.\n')

binary_parition_index(norm.cdf,1.5,6)
```

```
111011.
```

Solution 9.4 *Polya tree sampling.* The following function `polya_tree_sample` is one possible Python implementation. For a user-chosen depth m, the code samples the bin probabilities $\mathbb{P}(B_e)$ for each set $B_e \in \pi_m$, and then estimates the density at the mid point of the bin to be equal to the average density of the bin.

```python
#! /usr/bin/env python
# polya_tree_sample.py

import numpy as np
import matplotlib.pyplot as plt
from scipy.stats import norm

def polya_tree_sample(a,F_0_inv=norm.ppf,m=9,seed=0):
    gen = np.random.default_rng(seed=seed)
    p = np.ones(2**m)
    for j in range(m,0,-1):
        a_j = a*j**2
        i = 0
        for k in range(2**(j-1)):
            b = gen.beta(a_j,a_j)
            for l in range(2**(m-j)):
                p[i+l]          *= b
                p[i+l+2**(m-j)] *= 1-b
            i += 2**(m-j+1)

    x = np.array([F_0_inv((z+1.0)/2**m) for z in range(2**m-1)])
    return((x[1:]+x[:-1])/2, p[1:-1]/np.diff(x))

plt.plot(*polya_tree_sample(1))
plt.plot(*polya_tree_sample(100))
plt.plot(*polya_tree_sample(10000))
plt.autoscale(enable=True, axis='x', tight=True)
plt.gcf().canvas.manager.set_window_title('Polya tree samples')
plt.show()
```

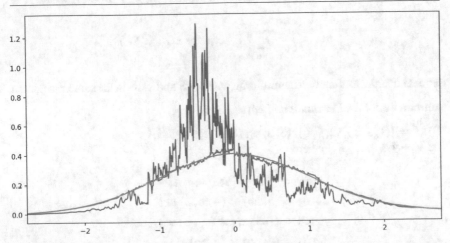

For larger values of α, the sampled densities more closely resemble the standard normal base measure density.

Solution 10.1 *Gaussian process closure.* For any $\mathbf{x} = (x_1, \ldots, x_n)$, independently
$$f(\mathbf{x}) - m(\mathbf{x}) \sim \mathrm{Normal}_n(0, K(\mathbf{x}, \mathbf{x})) \quad \text{and} \quad m(\mathbf{x}) \sim \mathrm{Normal}_n(m_0(\mathbf{x}), K_0(\mathbf{x}, \mathbf{x})),$$
where $K_0(\mathbf{x}, \mathbf{x})$ is the corresponding covariance matrix (10.3) for the kernel k_0. Since the sum of two independent normal distributions is again normal,

$$f(x) \sim \mathrm{Normal}_n(m_0(\mathbf{x}), K_0(\mathbf{x}, \mathbf{x}) + K(\mathbf{x}, \mathbf{x}))$$

and hence $f \sim \mathrm{GP}(m_0, k + k_0)$.

Solution 10.2 *Linear model as a Gaussian process.*

$$\beta \sim \mathrm{Normal}_p(0, \sigma^2\lambda^{-1}I_p) \implies X\beta \sim \mathrm{Normal}_n(0, \sigma^2\lambda^{-1}XX^\mathsf{T}).$$

It therefore follows that this regression function can be written a Gaussian process $f \sim GP(0, k)$ where the covariance kernel k is the dot product

$$k(x, x') = \sigma^2 \lambda^{-1} x \cdot x'.$$

Solution 10.3 *Spline regression as a Gaussian process.* It follows from Exercise 10.2 that $f \sim GP(0, v \cdot b(\cdot, \cdot))$ where

$$b(x, x') = 1 + \sum_{j=1}^{d} (xx')^j + \sum_{j=1}^{m} \{(x - \tau_j)_+ (x' - \tau_j)_+\}^d.$$

Solution 10.4 *Normal changepoint model as a Gaussian process.* The changepoint model is equivalent to a zero-mean Gaussian process $GP(0, v \cdot b(\cdot, \cdot))$ where

$$b(x, x') = \sum_{j=0}^{m} \mathbb{1}_{[\tau_j, \tau_{j+1})}(x) \cdot \mathbb{1}_{[\tau_j, \tau_{j+1})}(x')$$

defines an indicator function determining whether x and x' lie in the same τ-segment.

Solution 10.5 *CART notation and partition.*

(i) $T = \{(1, 2, a), (2, 1, b), (3, 3, c), (6, 1, d)\}$.
(ii) $\pi = \{B_1, \ldots, B_5\}$ where

$$B_1 = (-\infty, b] \times (-\infty, a] \times \mathbb{R}$$
$$B_2 = (b, \infty) \times (-\infty, a] \times \mathbb{R}$$
$$B_3 = (-\infty, d] \times (a, \infty) \times (-\infty, c]$$
$$B_4 = (d, \infty) \times (a, \infty) \times (-\infty, c]$$
$$B_5 = \mathbb{R} \times (a, \infty) \times (c, \infty).$$

Solution 11.1 *Mixture of normals full conditionals.* Let $n_{j,-i} = \sum_{i' \neq i} \mathbb{1}_{\{j\}}(z_i)$ be the number of data points aside from x_i currently allocated to cluster j. Similarly let $\dot{x}_{j,-i} = \sum_{i' \neq i; z_{i'} = j} x_i$ and $\ddot{x}_{j,-i} = \sum_{i' \neq i; z_{i'} = j} x_i^2$. Then for $j = 1, \ldots, m$,

$$p(z_i = j \mid \mathbf{z}_{-i}, \mathbf{x}) \propto \frac{(\alpha_j + n_{j,-i}) \Gamma(a + (n_{j,-i} + 1)/2) (\lambda + n_{j,-i})^{\frac{1}{2}}}{\Gamma(a + n_{j,-i}/2) (\lambda + n_{j,-i} + 1)^{\frac{1}{2}}}$$
$$\times \frac{[b + \frac{1}{2}\{\ddot{x}_{j,-i} + x_i^2 - (\dot{x}_{j,-i} + x_i)^2/(n_j + 1 + \lambda)\}]^{a + \frac{n_{j,-i}+1}{2}}}{[b + \frac{1}{2}\{\ddot{x}_{j,-i} - \dot{x}_{j,-i}^2/(n_j + \lambda)\}]^{a + \frac{n_{j,-i}}{2}}}.$$

Solution 11.2 *Gibbs sampling mixture of normals.* The following code is one possible Python implementation. The hyperparameters are chosen to be $\alpha = 0.1$, $a = 0.1$, $b = 0.1$, $\lambda = 1$.

```python
#! /usr/bin/env python
## gibbs_sampling_normal_mixture.py

import numpy as np
import matplotlib.pyplot as plt
from scipy.special import gammaln

def full_con_p(x,sx,sxx,n,al,a,b,la):
    a_star = a+.5*n
    lp = np.log(al+n) + gammaln(a_star) - gammaln(a_star+.5)
    lp += .5*np.log(1-1/(la+n+1))
    lp += a_star*np.log(b+.5*(sxx - sx**2/(la+n)))
    lp -= (a_star+.5)*np.log(b+.5*(sxx+x**2 - (sx+x)**2/(la+n+1)))
    return(np.exp(lp))

def gibbs_sampling(x,m=2,mu=1,al=.1,a=.1,b=.1,la=1,M=100,seed=0):
    gen = np.random.default_rng(seed=seed)
    n = len(x)
    ns = np.array([int(n/m) for _ in range(m-1)]+[n-(m-1)*int(n/m)])
    z = np.repeat(range(m),ns)
    sx = [sum([x[i] for i in range(n) if z[i]==j]) for j in range(m)]
    sxx = [sum([x[i]**2 for i in range(n) if z[i]==j]) for j in range(m)]
    pz = np.empty(m)
    for _ in range(M):
        for i in range(n): #loop through sample allocations
            for j in range(m):
                if z[i]==j:
                    ns[j] -= 1
                    sx[j] -= x[i]
                    sxx[j] -= x[i]**2
                pz[j]=full_con_p(x[i],sx[j],sxx[j],ns[j],al,a,b,la)

            z[i] = gen.choice(m,p=pz/sum(pz))
            ns[z[i]] += 1
            sx[z[i]] += x[i]
            sxx[z[i]] += x[i]**2

    print(", ".join(map(str,ns/n)))
    print(", ".join(map(str,np.array(sx)/ns)))

def simulate_beta_mixture(n, beta_pars, probs):
    gen = np.random.default_rng(seed=0)
    z = gen.choice(len(probs), n, p=probs)
    return(np.array([gen.beta(*(beta_pars[z_i])) for z_i in z]))

x = np.sort(simulate_beta_mixture(10000,[[20,10],[2,3]],[0.3,0.7]))
gibbs_sampling(x,2)
```

```
0.4869, 0.5131
0.3012809848370137, 0.6509450466163792
```

Solution 11.3 *Latent factor linear model.*

(i) $y \mid \sigma, X, Z \sim \text{Normal}_n(0, \sigma^2(XVX^\top + ZUZ^\top + I_n))$.
(ii) $y \mid X, Z \sim \text{St}_n(2a, 0, b(XVX^\top + ZUZ^\top + I_n)/a)$.

Solution 11.4 *Latent factor linear model code.* From Exercise 11.4 with $V = v\, I_p$ and $U = u\, I_q$,

$$y \mid X, Z \sim \text{St}_n(2a, 0, b(vXX^\top + uZZ^\top + I_n)/a).$$

The following Stan code is one possible implementation.

```
// linear_model_latent_factors.stan
data {
    int<lower=0> n; // number of observations
    int<lower=1> p; // number of grid points
    vector[n] y; // responses
    matrix[n, p] X; // factor matrix
    int<lower=1> q; // number of latent factors
    real<lower=0> u;
    real<lower=0> v;
    real<lower=0> a;
    real<lower=0> b;
}

transformed data {
    matrix[p, p] vXXTI = v * X * X' + identity_matrix(n);
}
parameters {
    matrix[n, q] Z; // latent factor matrix
}
model {
    matrix[n, n] S = u * Z * Z' + vXXTI;
    y ~ multi_student_t(2*a,rep_vector(0, n), b * S / a);
}
```

Glossary

$\mathbb{P}$	probability
$\mathbb{E}$	expectation
$\mathbb{V}$	variance
$:=$	definition
$\propto$	proportional to
$\rightarrow$	converges to
$\sim$	distributed as
$\implies$	implies
$\iff$	equivalent to
$\perp\!\!\!\perp$	independent
$\cdot$	dot product
$\mathbb{R}$	real numbers
$\mathbb{N}$	natural numbers, starting at zero
$\overline{A}$	set complement, $\{x \mid x \notin A\}$
$\mathbb{1}_A(x)$	indicator, 1 if $x \in A$, 0 otherwise
I_n	$n \times n$ identity matrix
B^T	transpose of matrix B
$|B|$	Determinant of matrix B
$||v||$	Euclidean norm of vector v
$|x|$	Absolute value of real value x

© The Editor(s) (if applicable) and The Author(s), under exclusive license to Springer
Nature Switzerland AG 2021
N. Heard, *An Introduction to Bayesian Inference, Methods and Computation*,
https://doi.org/10.1007/978-3-030-82808-0

References

Amaral Turkman MA, Paulino CD, Müller P (2019) Computational Bayesian statistics: an introduction. Institute of Mathematical Statistics Textbooks. Cambridge University Press, Cambridge

Barber D (2012) Bayesian reasoning and machine learning. Cambridge University Press, Cambridge

Berger JO, Guglielmi A (2001) Bayesian and conditional frequentist testing of a parametric model versus nonparametric alternatives. J Am Stat Assoc 96(453):174–184

Bernardo JM, Smith AFM (1994) Bayesian theory. Wiley Series in Probability & Statistics. Wiley

Betancourt M (2017) A conceptual introduction to Hamiltonian Monte Carlo. arXiv: Methodology

Bhattacharya A, Dunson DB (2011) Sparse Bayesian infinite factor models. Biometrika, pp 291–306

Bishop CM (2006) Pattern recognition and machine learning. Springer, Berlin

Blei DM, Ng AY, Jordan MI (2003) Latent Dirichlet allocation. J Mach Learn Res 3:993–1022

Blei DM, Kucukelbir A, McAuliffe JD (2017) Variational inference: a review for statisticians. J Am Stat Assoc 112(518):859–877

Chipman HA, George EI, McCulloch RE (1998) Bayesian CART model search. J Am Stat Assoc 93(443):935–948

Chipman HA, George EI, McCulloch RE (2010) Bart: Bayesian additive regression trees. Ann Appl Stat 4(1):266–298

de Finetti B (2014) Theory of probability: a critical introductory treatment. Wiley Series in Probability and Statistics, Wiley

Denison DGT, Mallick BK, Smith AFM (1998) A Bayesian CART algorithm. Biometrika 85(2):363–377

Denison DGT, Holmes CC, Mallick BK, Smith AFM (2002) Bayesian methods for nonlinear classification and regression. Wiley Series in Probability and Statistics, Wiley

Doucet A, de Freitas N, Gordon N (2001) An introduction to sequential Monte Carlo methods. Springer, New York, pp 3–14

Ferguson TS (1973) A Bayesian analysis of some nonparametric problems. Ann Stat 1(2):209–230

Fraley C, Raftery AE (2002) Model-based clustering, discriminant analysis, and density estimation. J Am Stat Assoc 97(458):611–631

Gelman A, Hennig C (2017) Beyond subjective and objective in statistics. J R Stat Soc Ser A (Statistics in Society) 180(4):967–1033

Gelman A, Carlin J, Stern H, Dunson D, Vehtari A, Rubin D (2013) Bayesian data analysis, 3rd edn. Chapman & Hall/CRC Texts in Statistical Science. Taylor & Francis

Green PJ (1995) Reversible jump Markov chain Monte Carlo computation and Bayesian model determination. Biometrika 82(4):711–732

© The Editor(s) (if applicable) and The Author(s), under exclusive license to Springer   165
Nature Switzerland AG 2021
N. Heard, *An Introduction to Bayesian Inference, Methods and Computation*,
https://doi.org/10.1007/978-3-030-82808-0

Green PJ, Silverman BW (1994) Nonparametric regression and generalized linear models: a roughness penalty approach. Chapman and Hall, United Kingdom

Heard NA (2020) Naheard/changepoints: Bayesian changepoint modelling code. https://doi.org/10.5281/zenodo.4158489

Hoffman MD, Gelman A (2014) The No-U-Turn sampler: adaptively setting path lengths in Hamiltonian Monte Carlo. J Mach Learn Res 15(47):1593–1623

Holmes CC, Denison DGT, Ray S, Mallick BK (2005) Bayesian prediction via partitioning. J Comput Graph Stat 14(4):811–830

Jeffreys H (1961) Theory of probability, 3rd edn. Englan, Oxford

Kass RE, Raftery AE (1995) Bayes factors. J Am Stat Assoc 90(430):773–795

Kucukelbir A, Tran D, Ranganath R, Gelman A, Blei DM (2017) Automatic differentiation variational inference. J Mach Learn Res 18(14):1–45

Lau JW, Green PJ (2007) Bayesian model-based clustering procedures. J Comput Graph Stat 16(3):526–558

Lavine M (1992) Some aspects of Polya tree distributions for statistical modelling. Ann Stat 20(3):1222–1235

Leonard T (1973) A Bayesian method for histograms. Biometrika 60(2):297–308

Mauldin RD, Sudderth WD, Williams SC (1992) Polya trees and random distributions. Ann Stat 20(3):1203–1221

Minka TP (2001) Expectation propagation for approximate Bayesian inference. In: Proceedings of the seventeenth conference on uncertainty in artificial intelligence, UAI'01. Morgan Kaufmann Publishers Inc., pp 362–369

Neal RM (2011) Mcmc using hamiltonian dynamics. In: Handbook of Markov Chain Monte Carlo. Chapman and Hall/CRC, pp 113–162

Oh M-S, Raftery AE (2007) Model-based clustering with dissimilarities: a Bayesian approach. J Comput Graph Stat 16(3):559–585

Rasmussen CE, Williams CKI (2005) Gaussian processes for machine learning (Adaptive Computation and Machine Learning). The MIT Press

Richardson S, Green PJ (1997) On Bayesian analysis of mixtures with an unknown number of components (with discussion). J R Stat Soc: Ser B (Statistical Methodology) 59(4):731–792

Roberts GO, Rosenthal JS (2004) General state space Markov chains and MCMC algorithms. Probab Surv 1:20–71

Roberts GO, Gelman A, Gilks WR (1997) Weak convergence and optimal scaling of random walk Metropolis algorithms. Ann Appl Probab 7(1):110–120

Rue H, Martino S, Chopin N (2009) Approximate Bayesian inference for latent Gaussian models by using integrated nested Laplace approximations. J R Stat Soc: Ser B (Statistical Methodology) 71(2):319–392

Schwarz G (1978) Estimating the dimension of a model. Ann Stat 6(2):461–464

Sethuraman J (1994) A constructive definition of Dirichlet priors. Statistica Sinica 4(2):639–650

Teh YW (2010) Dirichlet processes. In: Encyclopedia of Machine Learning. Springer, Berlin

Teh YW, Jordan MI, Beal MJ, Blei DM (2006) Hierarchical Dirichlet processes. J Am Stat Assoc 101(476):1566–1581

Tierney L, Kadane JB (1986) Accurate approximations for posterior moments and marginal densities. J Am Stat Assoc 81(393):82–86

Wang C, Paisley J, Blei D (2011) Online variational inference for the hierarchical Dirichlet process. In: Proceedings of the fourteenth international conference on artificial intelligence and statistics, vol 15. JMLR Workshop and Conference Proceedings, pp 752–760

Zellner A (1986) On assessing prior distributions and Bayesian regression analysis with g prior distributions. Elsevier, New York, pp 233–243

Index

© The Editor(s) (if applicable) and The Author(s), under exclusive license to Springer
Nature Switzerland AG 2021
N. Heard, *An Introduction to Bayesian Inference, Methods and Computation*,
https://doi.org/10.1007/978-3-030-82808-0

Printed in the United States
by Baker & Taylor Publisher Services